Digital Transformation towards Smart Steel Manufacturing

"The changes presented in the book are very important for development of steelworks in the era of Industry 4.0. The book presents the results of direct research that was realized in the Polish steel market. In my opinion this book contributes to the management science studies and may also be useful for practical business."

—Prof. Ryszard Barcik, *Katowice Business University, Poland*

"The book presents changes in the Polish steel industry towards smart manufacturing. I think that the presented results of research can be a good case study to advanced analysis of industrial transformation in other countries. At the era of Industry 4.0, I recommend this book because its topic is current and important for industries."

—Prof. Witold Biały, *KOMAG Institute of Mining Technology, Poland*

"I would like to recommend this book about industrial digital changes in the strongly promoted Industry 4.0 concept. The author Bożena Gajdzik did own direct research in the segment of the Polish steel sector. The results of this research show that steel mills are moving towards smart steel manufacturing, so the Industry 4.0 concept is important for industries, it is worth further promoting."

—Prof. Wojciech J. Cynarski, *Full Professor, University of Rzeszow, Rzeszów, Poland*

"The monograph "Digital Transformation Towards Smart Steel Manufacturing" covers the digitalization of steelworks processes with an emphasis on new trends and an ecological approach. The book focuses on analyses of the Polish steel industry, which is one of the largest in Europe. It is important to note that, in general, there is a lack of publications dealing with digitization and Industry 4.0 technologies in continuous manufacturing processes."

—Dr. Roman Danel, *Member of Czech Society for System Integration and Researcher at VSTE—Institute of Technology and Business in České Budějovice, Czech Republic and VSB—Technical University of Ostrava, Faculty of Economics, Department of System Engineering and Informatics, Czech Republic*

"The transformation of Katowice after the system change in 1989 is remarkable. Katowice, known as a city producing coal and steal during the communist era, has become a sophisticated modern city, with a noteworthy steal industry. Previously it had polluted environment, but thanks to Industry 4.0 and digitalization, it has become an environmentally friendly and smart industry. The author describes this transformation vividly."

—Yoji Koyama, *Professor Emeritus, Niigata University, Japan*

"The book presents the results of original and innovative research carried out in Polish steel sector, regarding the transformation of this sector strongly influenced by both the Industry 4.0 technology and the climate transformation (Net Zero 2050 Strategy). I definitely recommend this book due to its very high scientific level, comprehensive and innovative approach to the issues considered."

—Andrzej W. Kozina, *Associate Professor of Management, Cracow University of Economics, Department of Public Policies, Poland*

"This work is a timely contribution to the energy-efficient and green digitalization of steel industries. It emphasizes smart manufacturing and supply chain processes in the Fourth Industrial Revolution, enhanced by Industry 5.0 technologies. The book is valuable for theoreticians and practitioners since it is based on scholarly literature and is rounded off with a comprehensive survey among Polish practitioners."

—John Andrew van der Poll, *Research Professor, Digital Transformation and Supply Chain Management, Graduate School of Business Leadership, University of South Africa, South Africa*

"I recommend the book *Digital Transformation towards Smart Steel Manufacturing. A Framwork for Steel Enterprises in Poland* that provides a structured approach to how traditional industries can evolve in the digital age. The book offers a fascinating look at how Industry 4.0 principles are transforming the steel industry, particularly in Poland. It highlights key trends in digitalization, making it relevant for both academia and businesses. The book provides a practical framework for companies aiming to modernize, making it a great read for those exploring the future of industrial manufacturing."

—Hana Štverková, *Associate Professor at VSB —Technical University of Ostrava, Czech Republic*

"Professor Gajdzik's "Digital Transformation towards Smart Steel Manufacturing" offers a concise yet comprehensive analysis of Industry 4.0 in Poland's steel sector, based on research involving 253 companies. The monograph effectively combines theory and practice, highlighting the crucial role of decarbonization in smart steel manufacturing. In my opinion this is a valuable resource for researchers and industry professionals."

—Wieslaw Urban, *Professor of Management, Bialystok University of Technology, Poland*

"*Digital Transformation towards Smart Steel Manufacturing. A Framework for Steel Enterprises in Poland* offers a comprehensive exploration of the intersection between digital technologies and the steel industry, specifically in Poland. By combining theoretical insights with direct research, the book highlights the impact of Industry 4.0 and 5.0 on steel manufacturing and decarbonisation. It is an essential read for professionals and researchers looking to understand how automation, digitisation, and sustainability are transforming the steel sector globally and regionally."

—Robert Wolny, *Associate Professor at the University of Economics in Katowice, Poland*

"The steel industry has a direct impact on the competitiveness and resilience of the European Union industry. This strategic sector is facing many challenges today. This book helps to analyze and map the changes in the steel industry due to digital transformation and presents a unique smart model framework supported by direct research in enterprises."

—Frantisek Zapletal, *Professor, Head of the Department of Systems Engineering and Informatics, VŠB—Technical University of Ostrava, Czech Republic*

Bożena Gajdzik

Digital Transformation towards Smart Steel Manufacturing

A Framework for Steel Enterprises in Poland

palgrave
macmillan

Bożena Gajdzik
Silesian University of Technology
Gliwice, Poland

ISBN 978-3-031-77755-4 ISBN 978-3-031-77756-1 (eBook)
https://doi.org/10.1007/978-3-031-77756-1

This Palgrave Macmillan imprint is published by the registered company Springer Nature Switzerland AG
The registered company address is: Gewerbestrasse 11, 6330 Cham, Switzerland

Introduction

In the information age and open innovations, the functioning of businesses is significantly influenced by digital and intelligent technologies. The implementation of information and communication technologies (ICTs), together with the automation and robotisation of routine operations in processes, is transforming businesses. In the strongly promoted, for more than a decade, the concept of "Industry 4.0", more companies are participants of "smart" solutions. The term "smart" is synonymous with agility, flexibility, intelligence and other terms. We use the term when we want to create smart manufacturing and smart supply chain.

Smart technologies together with the Internet of Things (IoT), in recent years, there are global mega trends based on radical changes in manufacturing and logistics processes. For several years, an important objective of governments worldwide has been to promote the idea of smart manufacturing, which is intended to increase business efficiency and reduce operating costs, while building new quality products and services (new materials, smart products, online services). Digital technologies, automation and robotisation, which were initiated in the Third Revolution, have become part of intelligent steelmaking systems in the Fourth. The digital transformation, in recent years, is clearly moving towards artificial intelligence (AI) and the merging of cyber solutions with physical objects (CPSs). Industries, including steel study, changed their processes towards smart. In sustainable economies, smart technologies must be energy efficient, green, responsible and resident. Transformation changes are strongly involved in global energy policy and others. Digitisation is a component of global megatrends.

This book presents the impact of digital transformation (DT) and other megatrends on the changes of steel companies towards smart steel manufacturing. The book focuses on the digitisation of processes, the automation and robotisation of operations in steel enterprises in Poland. Technological innovations are presented in steel industry. Changes are taking place within steel mills as well as within supply chains (B2B). An essential part of the book is the results of direct research, which dealt with the transformation of enterprises to smart steel manufacturing. The research was done on the Polish steel market, which is an open market with foreign capital. The sample of 253 consisted of representatives of steel enterprises in Poland that served both domestic and foreign markets in this country. An essential part of the book is the results of direct research, which dealt with the transformation of enterprises to smart steel manufacturing. The research was done in the Polish steel market, which is an open market with foreign capital. The sample of 253 consisted of representatives of steel companies that served both domestic and foreign markets. And while thc research was strictly about the implementation of the pillars of Industry 4.0, the results were placed on the strategic transformation of the steel industry, which is strongly influenced by the political requirements of reducing CO_2 emissions ("deep decarbonisation").

The aim of the monograph is to present the frameworks of the processof digital transformation (DT) in steel enterprises in Poland based on pillars of Industry 4.0. The model framework was called "smart steel manufacturing". This model was described on the basis on the direct research in the segment of Polish steel enterprises.

The added value of the monograph is its own direct research conducted in steel market in Poland. The research focused on implementing of the pillars of Industry 4.0 towards smart steel manufacturing in this market.

The monograph consists of theoretical and practical parts. The first and second chapters present results of literature review strongly involved with the analysis of the steel market. The third chapter is about the Polish steel market. The fourth chapter consists of the results of own direct research (253 respondents from the Polish steel market). The last chapter describes components of digital transformation in steel surrounding, which the author calls "deep transformation" for importance analysed changes of Industry 4.0 (enhanced by Industry 5.0), as well as decarbonisation of steel industry.

Contents

1 **Megatrends Driving Modern Industrial Transformation**	1
1.1 Challenges of industrial transformation	2
1.2 Industrial Automation and robotisation	4
1.3 Trends in Industrial Digitisation	10
1.4 Technological pillars of Industry 4.0	11
1.5 Strategic Directions of Industry 5.0	13
References	19
2 **Steel Industry in the Era of Digitalisation and Climate Change**	27
2.1 Digitisation of Processes in Steel Industry	28
2.2 Background of Smart Steel Manufacturing	33
2.3 Towards Smart Steel Supply Chain	35
2.4 Decarbonisation of Steelmaking Process Towards "Green Steel"	41
References	48
3 **The Polish Steel Industry in the Context of Industrial Transformation**	55
3.1 Polish Steel Market	55
3.2 Polish Steel Industry in European Steel Market	69
References	76

4 Transitioning Towards Smart Steel Manufacturing: Results from Research in Poland 79
4.1 Methodology of Research 80
4.2 Results of Direct Research 84
4.3 Research Conclusions 118
4.4 Limitation of the Research and Future Study 121
References 122

5 A Model Framework for Smart Steel Transformation 125
5.1 Model Framework of Smart Steel Manufacturing 126
5.2 Key Directions of Smart Steel Transformation 133
References 135

6 Summary 139
6.1 Key Drivers for Digital Transformation in Steel Manufacturing 139
6.2 Technological Pillars of Smart Steel Manufacturing 140
6.3 Challenges to Digital Transformation in the Polish Steel Enterprises 141
6.4 Conclusion 141

Index 143

List of Figures

Fig. 1.1	The model of relation Industry 5.0 and Industry 4.0 with management levels. Source: own analysis	18
Fig. 3.1	Steel production in Poland from 1989 to 2022 (based on data of WorldSteel reports)	59
Fig. 3.2	Process of continuous steel production in Poland from 1989 to 2022. Source: Own elaboration based on data from Polish Steel Association in Katowice	60
Fig. 3.3	Steel production and steel consumption in Poland from 2005 to 2022. Source: Own elaboration based on data from Polish Steel Association in Katowice	61
Fig. 3.4	Consumers of steel products in Poland. Source: Own elaboration based on data from Polish Steel Association in Katowice	62
Fig. 3.5	Foreign trade of steel products in Poland from 2004 to 2022. Source: Own elaboration based on data from Polish Steel Association in Katowice	63
Fig. 3.6	Employment and production efficiency in the Polish steel market from 1989 to 2022. Source: Own elaboration based on data from Polish Steel Association in Katowice	65
Fig. 3.7	Energy resources in the Polish steel market. Own elaboration based on data from Polish Steel Association in Katowice	66

Fig. 3.8 Energy consumption per 1 tonne of steel in Poland from 2011 to 2021. Source: Own elaboration based on data from Statistics Poland (GUS) Warsaw, reports: Energy efficiency in Poland 67
Fig. 3.9 Expenditures on investments in steel industry in Poland from 1989 to 2022. Own elaboration based on data from Polish Steel Association in Katowice 68
Fig. 3.10 CO_2 emissions per tonne of steel in Poland from 2005 to 2022. Source: own elaboration based on KOBiZE report obtained from Polish Steel Association, Poland, Katowice 75
Fig. 4.1 The structure of research analysis. Source: own research 81
Fig. 4.2 Boxplot of Response Distribution and Median/Mean Comparison Across Questions with Tukey's HSD Test Results in the research area: RA_1. Source: own research 89
Fig. 4.3 Cluster dendrogram of response similarity in the research area: RA_1. Source: own research 91
Fig. 4.4 Distance matrix between questions with Manhattan distance in the research area: RA_1. Source: own research 92
Fig. 4.5 K-medoids cluster analysis in the research area: RA_1. Source: own research 94
Fig. 4.6 Discrepancy analysis in the research area: RA_1. Source: own research 96
Fig. 4.7 Scatter plot of horizontal and vertical analysis in the research area: RA_1. Source: own research 97
Fig. 4.8 Boxplot of Response Distribution and Median/Mean Comparison Across Questions with Tukey's HSD Test Results in the research area: RA_2. Source: own research 100
Fig. 4.9 Cluster dendrogram of response similarity in the research area: RA_2. Source: own research 101
Fig. 4.10 Distance matrix between questions with Manhattan distance in the research area: RA_2. Source: own analysis 103
Fig. 4.11 Boxplots for observed states in HMM. Source: own research 106
Fig. 4.12 Boxplot of Response Distribution and Median/Mean Comparison Across Questions with Tukey's HSD Test Results in the research area: RA_3. Source: own research 109
Fig. 4.13 Cluster dendrogram of response similarity in the research area: RA_3. Source: own research 110
Fig. 4.14 Distance matrix between questions with Manhattan distance in the research area: RA_3. Source: own research 111

Fig. 4.15 Discrepancy analysis in the research area: RA_3. Source: own research 113
Fig. 4.16 Opportunities and threats of smart industrial transformation in the Polish steel industry. Source: own research 115
Fig. 4.17 Boxplot of Response Distribution and Median/Mean Comparison Across Questions with Tukey's HSD Test Results in the research area:RA_4. Source: own research 116
Fig. 4.18 Planned directions of smart transformation in steel industry. Source: own research 119
Fig. 4.19 Changes on the timeline: progress of changes in the smart transformation of the Polish steel industry. Source: own research 119
Fig. 5.1 Model framework of steel industry transformation towards "smart" and "green' steel. Source: own elaboration 128
Fig. 5.2 Model framework of steel transformation in the era of Industry 4.0. Source: own elaboration 130

List of Tables

Table 1.1	Digital technologies in enterprise processes	8
Table 1.2	Categories of the Industry 4.0 pillars	14
Table 2.1	Levels of used computer systems in steel mills	29
Table 2.2	Levels of digital influence on steel processes	30
Table 2.3	Layers of smart steel manufacturing	35
Table 2.4	Solutions of vertical integration in steel industry	36
Table 2.5	Results of used new technologies in supply chains	39
Table 2.6	Decarbonisation technologies in the Green Steel for Europe (GREENSTEEL)	45
Table 2.7	Key green steelmaking initiatives	47
Table 3.1	Main steel products in Poland (million tonnes)	63
Table 3.2	Key realised investments in the Polish steel industry	70
Table 3.3	Steel production in Central and Eastern European countries	72
Table 3.4	Steel products total: Poland vs. the EU	73
Table 4.1	Matrix of research structure: survey	81
Table 4.2	Structure of response shares in the RA_1 based on the questions: Q_1 and Q_2	85
Table 4.3	Results of CLMM analysis using R computer program	86
Table 4.4	Post hoc Tukey test comparisons using “R” computer program	87
Table 4.5	Score Method Clusters	94
Table 4.6	Answers of respondent 214 (medoid of the cluster analysis)	94

Table 4.7	Answers of respondent 221 (medoid of the cluster analysis)	95
Table 4.8	Structure of response shares in the RA_2 based on the questions: Q_3 and Q_4	98
Table 4.9	AI and BIC analysis using "R" computer program	104
Table 4.10	Structure of response shares in the RA_3 based on the questions: Q5 and Q6	107
Table 4.11	Structure of response shares in the RA_4 based on the questions: Q7 and Q8	114
Table 4.12	Structure of response shares in the RA_5 based on the questions: Q9 and Q10	118

CHAPTER 1

Megatrends Driving Modern Industrial Transformation

Abstract Modern industries are subject to a number of trends that affect businesses strongly. Those trends whose impact is significant are referred to as megatrends. Such megatrends include the concept of Industry 4.0, which has been strongly popularised since 2011. This new concept is described in nine pillars, which are the constructs of smart manufacturing, or smart factory. However, to build (create) smart solutions, it is first necessary to invest heavily in digitisation and automation, which have received significant support from robots in recent years in line with this direction of change. This chapter describes the challenges of industrial transformation in the following order, first automation of operations and robotisation, then digitisation of industry and finally the assumptions of Industry 4.0, which, after the COVID-19 pandemic, were reinforced by the strategic directions of Industry 5.0.

Keywords Industry 4.0 · Industry 5.0 · Industrial transformation · Automation · Robotisation

B. Gajdzik, *Digital Transformation towards Smart Steel Manufacturing*, https://doi.org/10.1007/978-3-031-77756-1_1

1.1 Challenges of industrial transformation

Industries have changed a great deal over the last quarter of a century. Current trends have emerged in many industries and economies and the dependence of businesses on government environmental policies has increased. Dynamic changes in economies and markets have forced companies to study change and business transformation. Small trends, with varying degrees of impact on business, have given way to large, major changes that are radically transforming businesses. Among the most significant determinants of the functioning of economies, societies and organisations today are: the globalisation of economies, the digitisation of various spheres of human and business activity, the acceleration of automation and robotisation of operations, smart technologies, smart products and services, the vertical and horizontal integration of business areas, quality orientation, resource savings, sustainability and the responsibility of businesses towards the environment and various stakeholder groups (Łapuńka & Marek-Kołodziej, 2017).

The development of industries is the result of successive industrial revolutions. The First Industrial Revolution dates back to 1784, when steam was applied to industry and the mechanical era of production occurred, the Second Revolution began in 1870, at which time industry was electrified and the Third Revolution began in 1969, as a result of the development of electronics and information and computer technology (IT) and manufacturing was increasingly automated, as for the date of the Fourth Industrial Revolution, it was difficult to establish, due to the flexible transition of digital technologies to smart with Internet of Things (IoT) and Internet of Everything (IoE) (Paprocki, 2016).

The development of digital technologies and the computerisation of processes have shaped the new economies known as the digital economies. Digitality has entered many industries and services. Entrepreneurs have proceeded to form strong corporations and serve more customers in the markets. Modern business knows no borders, with branches and subsidiaries of companies located in many countries around the world. Such changes also occurred in the steel sector and intensified at the turn of the century, when strong corporate groups such as ArcelorMittal were established (Gajdzik & Sroka, 2012). In the Fourth Industrial Revolution, global cooperation was made possible by the digitisation of processes. D. Tapscott (1999) called such collaboration "networked intelligence" (NI).

Such cooperation develops markets based on the idea of the concept of Industry 4.0.

The term "Industrie 4.0" was first used at the Hanover Fair in 2011, and in October 2012, a working group, led by Siegfried Dais of Robert Bosch GmbH, presented a set of implementation recommendations to the German federal government for industry using innovative technologies, referred to as "high" according to the title of the paper's entry: High Tech Strategy 2020. The group's final research report was presented on 8 April 2013 and included recommendations and suggested directions for change for the industry of the fourth revolution, whose technologies have become the pillars of Industry 4.0 (Hermann et al., 2015; Plattform Industrie 4.0, 2014).

The innovative technologies of the Fourth Industrial Revolution such as Big Data, cloud computing, IoT and IIoT, machine learning (ML) and artificial intelligence (AI), additive manufacturing with 3D printing, as well as others are strongly influencing the innovation in manufacturing companies. Modern technologies will create, in the future, smart factories (Erboz, 2017; Hermann et al., 2015; Kagermann et al., 2011; Schwab, 2016).

Changes in modern industries are the result of the introduction of innovative technologies into the operation of businesses that focus on the value derived from flexible chains. Innovative industries cannot exist without ICTs, Internet of Things (IoT), cloud computing and Big Data Analytics. Developing cyber-physical systems (CPS) control physical processes, creating virtual (digital) copies of the real world and making decentralised decisions, and through the Internet of Things, communicate and collaborate in real time and through the cloud, use a range of operational services (Schwab, 2016).

With changes in industrial technologies, governments in many countries are talking about the need to use innovative technologies to stop the warming of the climate. The participants of the twenty-first session in Paris (from 30 Nov to 11 Dec. 2015) in (The Paris Agreement, 2015) adopted an important climate goal, which is to reduce carbon emissions radically and rapidly and "net zero" by 2050. Deep decarbonisation, as the direction for achieving the net zero target was called, as well as other targets for reducing natural resource and energy consumption, are embedded in the global "Green Economy" programme (web of

UNEP). This programme is linked to sustainable consumption, sustainable production and resource efficiency in the life cycle of products (LCM) towards QLCA (Gajdzik et al, 2024).

The importance of technological progress for economies and industries is also enshrined in the assumptions of the next industrial development concept called "Industry 5.0" (Industry 5.0, European Union, Jan. 2021). The strategic directions of the concept are sustainability, the centrality of human factor and resilience. The EU publication on Industry 5.0 (2021) states that it is necessary to learn how to use "high" technologies for intensive environmental protection, human development and business resilience in the "lability" (Taleb, 2007)· Business must understand "anti-fragility" and can rebirth after extreme and random events (Taleb, 2012). Modern business will increasingly be faced with difficult to predict situations caused by unexpected events. Economies, industries and societies must change and become resilient and therefore must transform in line with global trends.

1.2 Industrial Automation and robotisation

Digitisation together with smart technologies, according to prominent researchers (Di Silvestre et al., 2018; Kagermann, 2015; Kagermann & Winter, 2018; Pahalson et al., 2021), are radically affecting economies, industries and societies almost worldwide. The digital transformation, of the Fourth Industrial Revolution, referred to as the "second wave" (Kagermann & Winter, 2018), has been included in the concept of Industry 4.0. Digitisation has been recognised as the basis for cyber-physical systems (CPSs).

Digitisation (i.e. the process of transforming analogue data into digital data sets), combined with the technologies attributed to the concept of Industry 4.0 (cloud computing, smart sensors, big data, 3D printing, artificial intelligence to support machine learning, computer-based vertical and horizontal business integration systems, virtual and augmented reality solutions), has opened up new opportunities to bring about changes in processes within companies and across supply chains. Many manufacturing industries are adopting information and computer technologies (ICTs) to improve the flow of processes and forms of communication with the markets (Erboz, 2017; Hermann et al., 2016). The modern technologies proposed for Industry 4.0 (I 4.0) can create conditions for additive manufacturing and smart manufacturing in companies in many industries of the world.

The digital changes can be used by industries in the processes of deep decarbonisation (COP21). The changes affect both the electricity sector and many others, which are highly energy-intensive, such as the steel sector. Climate policies, implemented at the global level, are transforming generation (green energy, zero-carbon electricity, nuclear energy, hydroelectric and wind energy etc.) and distribution (decentralisation of the energy market) and increasingly radically transforming generation technologies into key industries for economies (fuel, chemical, mining, metallurgical and many others) (Di Silvestre et al., 2018).

Digitisation is embedded in the concept of Industry 4.0. Digitisation is not one technology but many technologies. It is a process of convergence of the real and virtual worlds. This process is conducted non-stop. People's access to computers and the Internet has made digitisation widespread. In the modern world, advanced digital technologies are increasingly penetrating every aspect of lives and businesses.

Digitisation mixed with artificial intelligence (AI) is creating "smart digitisation" (Petrov, 2020). This digitisation needs Big Data, Big Data Analytics, IIoT and ML (Kowalczyk, 2017). The Internet of Everything (IoE) is a force of the industrial revolution. IoE and new functions of technologies create new forms of communication people-to-people (P2P), people-to-machine (P2M) and machine-to-machine (M2M). Both the Internet and ICTs are creating the hyper connectivity that is now creating smart factories and smart supply chains, with blockchain technologies (Shala et al., 2019).

Smart solutions are needed in the optimisation of production process and others. Dedicated or universal IT and industrial automation tools are used in manufacturing and logistic processes. Digitisation focuses both on core processes (manufacturing) and non-core processes (supply, distribution, administration, marketing and others) (Gajdzik & Štverková, 2023; Gajdzik & Wolniak, 2022). The support to the Internet of Things (IoT) and the Internet of Everything (IoE) in ubiquitous connectivity (hyper connectivity) are cloud-based applications and services, Big Data Analytics (BDA) and Big-Data-as-a-Service (BDaaS), full automation with autonomous robots, multi-channels and omni-channels, etc. (Pieriegud, 2016).

Digital and smart technologies are used to increase the productivity of processes. KPIs (Key Performance Indicators) are incorporated into the analysis of all processes in the manufacturing companies and their relations with the market. The widespread use of KPIs is easy because data from the devices is collected in real time, which at the same time allows them to be analysed in real time without much human effort (Davies, 2015; Heuser et al., 2008).

The importance of the digitisation of industries has been highlighted in EU programmes under the "DIGITAL EUROPE". The programmes focus on the creation of European Digital Innovation Hubs (EDIHs). The Digital Europe programme is implemented through multi-annual work programmes dedicated to the following areas of intervention, among others (Digital Transformation Monitor, 2017):

1. High Performance Computing (implemented under indirect management by the Euro HPC JU).
2. Big Data, AI, Cloud, Quantum Communication Infrastructure (a programme targeting advanced digital skills and deployment activities to make the best use of innovative technologies).
3. Cybersecurity (implemented under indirect management by the Cybersecurity Industrial, Technology and Research Competence Centre and the Network of National Coordination Centres).

Europe's digitisation programmes are based on constructive interaction and complementarity of different activities in economies. Several levels of digitisation can be distinguished in manufacturing companies (Vor dem Esche & Hennig-Thurau, 2014):

- digitisation of information (the digital storage and processing of information in business);
- digitisation of production according to principles of the pull strategy in the market and product personalisation (production organised for the customer);
- digitisation of communication at all levels of the organisation (company) and within supply chains, as well as between the company and its stakeholders;
- digitisation of distribution (the execution of purchases and the processes of delivering products to customers with the use of digital technologies);

- digitisation of other processes within the company and within the supply chain (vertical and horizontal integrations);
- digitisation of products and services (QR codes in products, online services, etc.).

Hyper connectivity via the Internet (and its dedicated forms) means that companies can collect and analyse processes in real time. Based on the analysis of Big Data, they can optimise processes and predict them. Processes are transformed into autonomous cyber-physical systems (CPSs). Cyber-physical systems (CPSs) create modern production lines with a greater number of autonomous robots. Production in CPSs is supported by transport and distribution. Manufacturing companies need efficient logistics (Barreto et al., 2017).

Technologies in CPSs can match resources to the processes. In many companies, the start of process digitisation was the system of Enterprise Resource Planning (ERP) (Hauke, 2017). Over time, other IT systems were installed to support resources processes, such as Warehouse Management Systems (WMS), Transportation Management Systems (TMS), Intelligent Transportation Systems (ITS), Production Planning and Scheduling Systems (AP—Advanced Protection Systems), Manufacturing Execution System (MES), Product Lifecycle Management (PLM) and systems for business analysis and management decision support, such as Business Intelligence (BI) (Mayer-Schonberger & Cukier, 2014). The availability of real-time big data, collected through systems, sensor networks, drone points and BI systems, can increase the efficiency of decision-making companies. According to Peraković et al. (2019), integrated systems and autonomous technologies support such functions: production planning and scheduling, operations monitoring, process controlling, reporting the work of production departments, analysis of equipment operation parameters, including reporting on machine efficiency with catalogues and archives technical documentation, management of internal transport and logistics, analysis of energy and resources intensity, heterogeneity and intermodality of IT communication networks and automation, warehouse management, maintenance services, asset management and registers (Counsell, 2009). The examples of digital technologies used in companies are presented in Table 1.1.

Table 1.1 Digital technologies in enterprise processes

Computer systems and technologies	*Main abilities of solutions*
Business Intelligence (BI)	BI has the ability to create automated and personalised processes, flexible adaptation of production to the unique needs of the customer and a wide range of proactive activities because the choice of proposed solutions was preceded by the analysis of the customer's behaviour, his purchasing history and sales experience
Integrated package of ERP system	The ERP package supports the management, primarily of the economic and financial sphere of activity of companies and their contacts with the business environment
Electronic payment systems and mobile technologies	Solutions such as e-invoicing, online services, company websites, social media, etc. help to stay connected with the consumers and analyse their behaviours
Automation of processes in the company/ robots management	Automation supports processes and eliminates problems without delay. Moreover, it reduces production time and increases the accuracy of operations. Automation reduces operational costs
Internet of Things (IoT) and Internet of Everything (IoE)	IoT and IoE create new opportunities for maximising the number of interactions of multiple participants with devices (P2M), and devices with each other (M2M)
Big Data	Big Data and digitisation create opportunities for the correlation of data sets (huge quantities) that are unrelated to each other and come from various sources. Big Data Analytics creates a whole new added value by combining data, finding data, establishing correlations, reusing data, expanding data, expanding the scope of analysis, organising disparate source data structures, penetrating datasets, etc

(continued)

Table 1.1 (continued)

Computer systems and technologies	*Main abilities of solutions*
Cloud computing	Cloud computing (provides the ability to collect and process data (internet communication), using the IT resources of the cloud provider. The company does not have to have an extensive IT architecture, and instead of buying systems and applications—companies pay a subscription for using ready-made or custom-made solutions available on the supplier's servers
Digital platforms	Platforms are used to manage objects and processes, in the technology of 2D imaging and modelling Virtual Reality 3D (employees equipped with special virtual goggles). Platforms have an "open" configuration and can combine several types of specialised software into one operating system, which makes it possible to build on them integrated simulation programs. Enterprise IT platforms have interactive, interdisciplinary and intuitive operating menus that allow learning to work on the platforms without using the user manual, based on intuitive selection operational functions
Mobile applications	Business digital websites and their mobile applications (images, maps, content, graphics, etc.), including interactive data transfer tools, navigation systems, non-public networks (extranets and others), are used in various forms of communication and cooperation with consumers and stakeholders
Sensors	Sensors are most often components of a larger system whose task is to capture signals from the surrounding environment, recognise them and record them. Sensors and transducers are used in process automation to detect changing measured quantities and sending output signals. Several types of sensors are used in companies (optical, electromagnetic, inductive, ultrasonic, capacitive, etc.)

Source: own elaboration based on Counsell (2009), Dörner and Edelman (2015), Chen et al. (2012)

1.3 Trends in Industrial Digitisation

The development of industries is strongly linked to the automation of operations in manufacturing processes and logistics. The industrial transformation in Industry 4.0 is based on the full automation of operations in factories. Although previous industrial revolutions were based on the automation of production: the First (late eighteenth century)—on steam engines; the Second (late nineteenth century)—on electricity and the creation of production lines; the Third (second half of twentieth century)—on computers and other electronic technologies, the current, the Fourth is the strongest because it combines Industrial Internet of Things (IIoT), autonomous robots, Big Data Analytics and other technological solutions.

The automated equipment in production, transport and warehouse is closely related to industrial and services robots. Industrial robots are devices that replace humans when performing repetitive manufacturing, transport and storage operations. In the standard ISO 8373 (Vocabulary), a robot is "a programmed actuated mechanism with a degree of autonomy to perform locomotion, manipulation or positioning".

In the past few years, robots have entered industries strongly, driven by the rapid growth of industrial robot markets. The World Robotics 2023 Industrial Robots and Service Robots report showed that 553,052 industrial robots were installed in factories worldwide in 2022, an increase of 5% compared to the previous year (Robotics based on World Robotics Report, 26 Sep. 2023).

According to the International Federation of Robotics (IFR), 73% of all new robots were installed in Asia, 15% in Europe and 10% in the Americas. The world record of 500,000 robots was in 2021 and 2022. In an optimistic scenario, the IFR expects to reach 700,000 robots in 2026. The global total of active industrial robots is 2 million units. The greatest growth in robotisation is in Asia, including China, while the leader, in Europe, is Germany, with a market share of 36% in the EU (2022). The industry using the most industrial robots is currently the automotive industry, which accounts for almost 40% of global sales. The second position belongs to the machinery and metals sector and the third to the electrical and electronics (IFR, 2023).

The progress of industrial robots is strongly linked to the concept of Industry 4.0, in which automation becomes more intelligent (Pieriegud,

2016). Industrial robots used in manufacturing achieve higher productivity than humans and lower labour costs. Robots installed in industry reduce the number of tasks performed by workers. Intelligent robots improve job quality and optimise processes. Robots perform most often manipulation operations with heavy objects (heavy-duty industrial robots) and precision operations (collaborative robot—cobot). In addition, robots serve customers and communicate with them (business robot—chatbot). Robots of the latest generation are intelligent devices with which you can achieve goals, tasks, and operations faster and more efficiently than humans. In the concept of Industry 4.0, robots are more autonomous, but this does not mean that they will eliminate humans from operates (Klimchik et al., 2017; McGarry et al., 2022; Romero et al., 2016a, 2016b; Zhu et al., 2020).

In the concept of Industry 4.0, robots and intelligent solutions can increase operational efficiency by the interaction between humans and machines (P2M) and between the machines (M2M) (Frey et al., 2013). Operators cooperate with machines and robots, e.g. explain and interpret the results of their work, help maintain them, teach them new operations, etc. (Romero et al., 2016a, 2016b). The human and robot collaboration in cyber-physical systems is called Human Cyber-Physical Systems (H-CPS) (Flores et al., 2020).

According to Gartner analyst, the trend towards human-to-machine (P2M) and machine-to-machine (M2M) process automation, i.e. machine intelligence applied to the work environment, will continue in the coming years (Groombridge, 2023). The artificial intelligence (AI) will allow the functioning of devices that acquire knowledge on their own and reach the goal set by humans in different ways. In improved robots will use technological advances such as machine learning (ML) and real-time cyber-physical connections to improve the accuracy of robots, enabling them to be more adaptable and autonomous (Drouot et al, 2018).

1.4 Technological pillars of Industry 4.0

Industrie 4.0—a German initiative of politicians and businesspeople—is a collection of many technological solutions, called technology components or technology pillars of high-tech strategies (Hermann et al., 2015; de Sousa Jabbour et al., 2018; Essakly et al., 2019). This concept, from 2011, refers to the digital transformation of industry but through the Industrial Internet of Things (IIoT), cyber-physical systems (CPSs) and smart factories (Hermann et al., 2015). "Industry 4.0" is sometimes also called: "Industrial Internet" or "Integrated Industry", "Smart Industry" or "Smart Manufacturing" because it has three levers, firstly, the development of the Internet of Everything, secondly, cyber-physical systems, thirdly, smart manufacturing and smart products (Evans & Annunziata, 2012). In the term: "Industrial Internet" the focus is on the use of the Internet in industries. The Internet integrates machines and physical devices with sensors and software on a network in business activities (Bungart, 2014; Industrial Internet Consortium, 2013). In the United States, the name "Industrial Internet" is used even more often than the name "Industry 4.0". The name "Integrated Industry" is used for networked systems in supply chains (Bürger & Tragl, 2014) and the names "Smart Industry" or "smart manufacturing (SM)" are used when one wants to highlight (expose) the latest virtual reality solutions and AI in manufacturing companies (Dais, 2014; Davis et al., 2012; Wiesmüller, 2014).

The concept of Industry 4.0 is based on pillars, described in key technologies. Researchers describe particular pillars together with their technological features. The description of Industry 4.0 pillars we can find in: Sinderman et al., 2016; Erboz, 2017, Burrell, 2019; Senn, 2019; Rüßmann et al., 2015; Gonçalves Machado et al., 2020). Senn (2019) Burrell (2019) and Gonçalves Machado et al. (2020) describe nine main pillars of Industry 4.0. A summary of the set of Industry 4.0 pillars is sorted out in the paper by Saniuk et al. (2020). The nine pillars of Industry 4.0 are:

- autonomous robots that can perform operations and collaborate with workers, making their work easier;
- virtual simulations that reduce machine and process setup time by very precisely reproducing real operations;
- vertical and horizontal integrations;
- Big Data and Big Data Analytics;
- Industrial Internet of Things (IoT) in which machines and products to interact with each other;
- cloud computing with software and other solutions that reduce the cost of data storage and system upgrades;
- additive manufacturing with 3D printing;
- augmented reality as a connection between a virtual world and a physical object;

- cybersecurity with protocols and procedures for control and identification, protecting systems against cyberattacks in real time.

Technologies in these pillars can be grouped into four categories: the first category is strictly technological, the second is strongly connected with processes, the third is related to products, machines and objects, and the fourth category is for consumers and users (Table 1.2). Companies use both rapid prototyping and smart manufacturing solutions. Users are interested in the development of IT systems that support business processes and make these processes autonomous and self-controlled. Products in processes are identified by codes. The manufacturing process are under the consumer. End users—consumers—want personalised products.

1.5 Strategic Directions of Industry 5.0

Industry 5.0 is the next vision of digital transformation and a strategy for economies that want to integrate industrial change more strongly with the business resilience, climate goals and human development in cyber-physical systems. In the concept, companies should properly consume technologies. In industrial development, there is still a preponderance of hardware solutions over software solutions, especially in processes and product maturity. Management systems of events should be an integral part of a machine (a collection of machines or a separate system). The entire event-based integration of systems in an enterprise should be an unlimited network of connections in the relationship "event - publication - reception of information". In the concept of Industry 5.0, when the amount of data from systems and sensors generated in a short time interval will be very large, it is necessary to take care of the quality of the data in the individual systems so that there is free machine learning and the full spectrum of possibilities of real artificial intelligence (Bernick, 2017).

The term Industry 5.0 was first used in 2017 by Esben H. Østergaard, CEO at REInvest Robotics. Esben H. Østergaard presented the human touch of robotics in cooperation between humans and intelligent production systems (Piątek, 2018). Two years later, the concept of Industry 5.0 was discussed by Aroop Zutshi, CEO of Frost & Sullivan. One of the first scientific publications was Birth of Industry 5.0: Making Sense

Table 1.2 Categories of the Industry 4.0 pillars

Category	*Components*
Technology	• Internet of Things • Cloud computing • Connectivity between devices and objects (M2M, Car2Car) • Industrial data networks (fieldbus systems) • Internet protocols (IPv6) • Industrial Ethernet • Virtual Private Network (VPN) • Wireless computer network standards (Wi-Fi) • Wireless short-range radio communication standards (Bluetooth, NFC) • Open, neutral, compatible data exchange standards (XML, Automation ML—AML) • Smart grids
Process	• Cyber-physical production systems (CPPS) • Data analysis and business analytics (BI) • IT systems for production planning (ERP, PLM, APS) • Computer-aided design (CAD, CAM, MES) • Architecture of service-based IT systems (SOA, SaaS, Simulation) • Adaptive programming • Augmented reality • Wearables • Collaborative robots • Artificial intelligence • Resilient factory • Flexible supply chains

(continued)

Table 1.2 (continued)

Category	*Components*
Product/object/machine	• Advanced materials • Two-dimensional code, radio tag (QR code, RFID), smart tag • Identification of vehicles by VIN codes (autoident) • Rapid manufacturing techniques, e.g. additive manufacturing, 3D printing, selective laser sintering • Cyber-physical systems (CPS) • Cybernetics (cybernetics) • Robotics and control program (NC code) • Predictive maintenance • Machine learning • Smart objects • Smart products
Consumer/user	• Product tailored to your needs and available on demand (customised and on demand) • Online communication • e-commerce • Participation in product design • Tracking of the manufacturing process • Transparency of data transfer—blockchain • Data integration • Support for the circular economy (recycling) • Sharing products

Source: own elaboration

of Big Data with Artificial Intelligence, "The Internet of Things" and Next-Generation Technology Policy (Özdemir & Hekim, 2018).

On the political level, the first comprehensive vision of the concept was created in January 2021 by the European Commission (Brussels, Industry 5.0 (…), EC, Jan. 2021). The document states that Industry 5.0 is based on three pillars (values) of development: "human-centric", "sustainable" and "resilient". However, this was not the first publication on Industry 5.0. It was preceded by a European Commission study from September 2020, a document by Julian Müller entitled: "Enabling Technologies for Industry 5.0. Results of a workshop with Europe's technology leaders". The study describes the technologies relevant to this phase and identifies social, governmental, political and economic challenges. Assumptions for Industry 5.0 were developed before 2020 and written in other documents but without explicitly defining the novelty of industrial development. Technologies in Industry 5.0 should contribute to a significant reduction in the consumption of natural raw materials, the introduction of new inputs into production, a reduction in the energy intensity of technologies, especially in energy-intensive industries (e. g. chemicals, metallurgical industries), a reduction in the carbon footprint of technologies (reducing carbon dioxide emissions), a reduction in energy from coal (black energy) in favour of renewable energy (green energy) (Majerník et al., 2022). According to the European Green Deal, industries must focus on greater energy efficiency and less carbon-intensive technologies and they must achieve carbon neutrality by 2050—a net zero strategy (the EU submitted its long-term strategy—https://unfccc.int/documents/210328—to the United Nations Framework Convention on Climate Change (UNFCCC) in March 2020: https://climate.ec.europa.eu/eu-action/climate-strategies-targets/2050-long-term-strategy_en).

In the concept of Industry 5.0, many technologies will take on the characteristics of "green" technologies, e.g. "Green IoT". Green means new aspects of sustainability, e.g. smart circular economy. Industry will invest more in carbon–neutral production technologies (Fraga-Lamas et al., 2021). The ongoing digital transformation will require refinement or even development of many regulations for their functioning, e.g. blockchain technologies, life cycle management, human-artificial cooperation. Industry 5.0 will increase productivity and operational efficiency, be environmentally friendly, reduce workplace accidents and shorten production cycles. Intelligent robots and systems will permeate supply chains and production halls faster than today. This will be enabled by the introduction of cheaper and highly efficient robots, made of advanced materials such as carbon fibre and lightweight but durable materials, powered by highly optimised battery packs, resistant to cyberattacks, with stronger data processing (i.e. big data and artificial intelligence) and a network of smart sensors. (Nahavandi, 2019).

Industry 5.0 will bring unprecedented challenges both in the field of humanisation of technological changes and sustainability (Grabowska et al., 2022). Technologies should help to assess potential threats. The systems will define connections between objects, which will be stored using blockchain technologies. Technologies from extended processes must enter supply chains, creating advanced chains for the entire product lifecycle (Al Mubarak, 2022; Dhiman & Nagar, 2022; Kaur et al., 2022; Leng et al., 2022; Singh et al., 2023; Szpringer, 2021). Humans will be at the centre of the blockchain (user-centric blockchain), and the entire blockchain will be strongly connected to the Internet of Things as BCoT (Blockchain of Things) and will be equipped with technologies and systems that can apply to chain "resilience" (De D. Karmakar et al., 2022; Espina-Romero et al., 2023; Yang et al., 2022; Sriman et al., 2022).

Furthermore, in the concept of Industry 5.0, cooperation between humans and intelligent machines will be more important. Employees in engineering and technical positions will follow (using IT support technologies) manufacturing processes (Nahavandi, 2019). The technical skills needed now include the operation of IT and computer systems, operation of intelligent technologies, management of devices distributed in the network. In addition to technical skills, analytical skills, conceptual skills and statistical knowledge are needed (Flores et al., 2020). According to Nahavandi (2019) a large number of jobs will be created in the intelligent systems arena, AI and robotics programming, maintenance, training, scheduling, repurposing and invention of a new breed of manufacturing robots. In addition, since repetitive tasks need not be performed by a human worker, it will allow for creativity in the work process to be boosted by encouraging everyone to innovatively use different forms of robots in the workplace.

The concept of Industry 5.0 is identified with Society 5.0 called an intelligent society because it uses information and computer technologies to collect data, transmit information and build knowledge (Nakanishi, 2019). Cyberspace makes it easier for society to update and share knowledge. New social and business chains are created in cyberspace. Society 5.0, compared to Society 4.0, which is called the information society, has a greater ability to integrate the real and digital (virtual) worlds (Nakanishi, 2019). The strong bond between these two worlds is used for human development as well as for solving social and economic problems within cities and regions, in the fields of energy management, disaster prevention, health care, agriculture and food, logistics and services (Carayannis et al., 2021).

To sum up the chapter, links between Industry 4.0 and Industry 5.0 were established at the enterprise management level. Human centred, sustainability, resilience—the three pillars of Industry 5.0 and the key technologies of Industry 4.0—the nine pillars of the concept have many connections, and although Industry 4.0 focuses mainly on the technical aspect and is being developed from an operational management perspective, and Industry 5.0 is being prepared from a strategic management perspective, these levels are connected. The Industry 4.0 view is the implementation of various new technical solutions into business entities especially in production to increase their level of innovativeness and effectiveness but the effectiveness on the tactical level must be connected with effectiveness on the strategic level of the company (Miyake, 2022; Sultan et al., 2021). Sometimes the company can implement new technological solutions but it is effectiveness will not increase in the long term due to a lack of strategical view, strategical thinking and strategical approach. Strategic level in the concept of Industry 5.0 is more visible than in Industry 4.0. Reliability between the technologies of Industry 4.0 and the strategic directions of Industry 5.0 will achieve promising efficiency, flawless production, minimum waste and customisable manufacturing (Gajdzik, 2023). In doing so, it will bring more people back to the workplace and improve the process efficiency. The tactical level is something between Industry 4.0 and Industry 5.0—on this level many activities are connected with both of those conceptions. Figure 1.1 presents the place of Industry 4.0 and Industry 5.0 in industrial transformation.

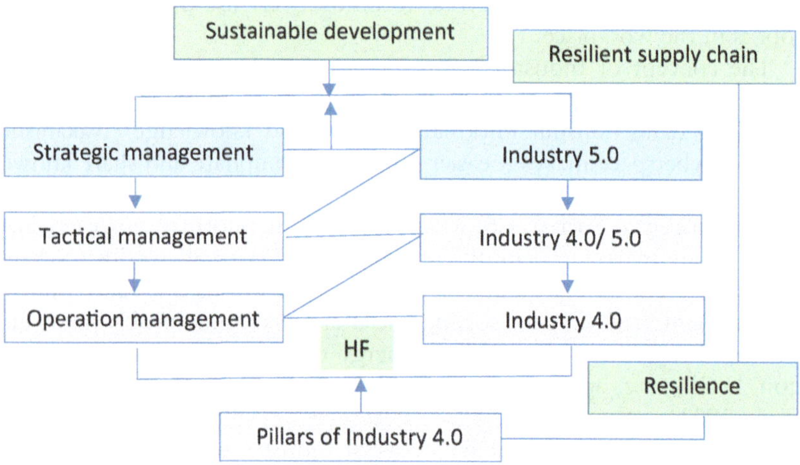

Fig. 1.1 The model of relation Industry 5.0 and Industry 4.0 with management levels. Source: own analysis

References

Al Mubarak, M. (2022). Sustainably developing in a digital world: Harnessing artificial intelligence to meet the imperatives of work-based learning in Industry 5.0. Development and Learning in Organizations.

Barreto, L., Amaral, A., & Pereira, T. (2017). Industry 4.0 implications in logistics; An overview. *Procedia Manufacturing, 13*, 1245–1252. https://doi.org/10.1016/j.promfg.2017.09.045

Bernick, M. (2017). After robots take over our jobs, then what? Retrieved from https://www.forbes.com/sites/michaelbernick/2017/04/11/after-robots-take-over-our-jobs-then-what

Bungart, S. (2014). Industrial Internet versus Industrie 4.0. Produktion – Technik und Wirtschaft für die deutsche Industrie. Retrieved March 15, 2019, from http://www.produktion.de/automatisierung/industrial-internet-versus-industrie-4-0/print

Burrell, D. (2019) Principles of Industry 4.0 and the 9 pillars. Retrieved from https://www.plextek.com/insights/insights-insights/industry-4-0-and-the-9-pillars/

Bürger, T., & Tragl, K. (2014). SPS-Automatisierung mit den Technologien der IT-Welt verbinden. In T. Bauernhansl, M. Ten Hompel, & B. Vogel-Heuser (Eds.), *Industrie 4.0 in Produktion, Automatisierung und Logistik. Anwendung, Technologien und Migration* (pp. 559–569).

Carayannis, E. G., Draper, J., & Bhaneja, B. (2021). Towards fusion energy in the Industry 5.0 and Society 5.0 context: Call for a global commission for urgent action on fusion energy. *Journal of the Knowledge Economy*. https://doi.org/10.1007/s13132-020-00695-5

Chen, H., Chiang, R., & Storey, V. (2012). Business intelligence and analytics: From big data to big impact. *MIS Quarterly, 4*, 1165–1188.

COP21: United Nations framework convention on climate change. In *21st Conference of the parties*. Paris Agreement, COP21, Le Bourget, France, 30 November 30–December 12, 2015, from cop21.gouv.fr

Counsell, J. (2009). Evaluation of 3D visualisation in the virtual environmental planning systems project. In R. Andrew & H. Alan (Eds.), *2nd International conference in visualization* (pp. 108–113). IEEE.

Davies R. (2015). Industry 4.0. Digitisation for productivity and growth, Briefing, ERPS. http://www.europarl.europa.eu/RegData/etudes/BRIE/2015/568337/EPRS_BRI.(2015).568337_EN.pdf

Dais, S. (2014). Industrie 4.0—Anstoß, Vision, Vorgehen. In: T. Bauernhansl, M. Ten Hompel, & B. Vogel-Heuser (Eds.), *2014: Industrie 4. 0 in Produktion, Automatisierung und Logistik. Anwendung, Technologien und Migration* (pp. 625–634).

Davis, J., Edgar, T., Porter, J., Bernaden, J., & Sarli, M. (2012). Smart manufacturing, manufacturing intelligence and demand-dynamic performance. *FOCAPO, 47*, 145–156.

de Sousa Jabbour, A. B. L., Jabbour, C. J. C., Foropon, C., & Filho, M. G. (2018). When titans meet—Can industry 4.0 revolutionise the environmentally-sustainable manufacturing wave? The role of critical success factors. *Technological Forecasting & Social Change, 132*, 18–25.

De, D., Karmakar, A., Banerjee, P. S., Bhattacharyya, S., & Rodrigues, J. J. P. C. (2022). BCoT: Introduction to Blockchain-based internet of things for Industry 5.0. *Lecture Notes on Data Engineering and Communications Technologies, 112*, 1–22.

Digital Transformation Monitor. (2017). *Key lessons from national Industry 4.0 policy Initiatives in Europe*. European Commission, European Union. More information online: https://www.digitaleurope.org/

Dhiman, G., & Nagar, A. K. (2022). Editorial: Blockchain-based 6G and industrial internet of things systems for Industry 4.0/5.0. *Expert Systems, 39*(10), e13162.

Di Silvestre, M. L., Favuzza, S., Sanseverino, E. R., & Zizzo, G. (2018). How decarbonisation, digitisation and decentralization are changing key power infrastructures. *Renewable and Sustainable Energy Reviews, 93*, 483–498. https://doi.org/10.1016/j.rser.2018.05.068

Dörner K., Edelman D. (2015). What 'digital' really means. McKinsey Digital July 2015. Retrieved December 20, 2023, from http://www.mckinsey.com/industries/high-tech/our-insights/what-digital-really-means

Drouot, A., Zhao, R., Irving, L., Sanderson, D., & Ratchev, S. (2018). Measurement assisted assembly for high accuracy aerospace manufacturing. In *Conference: 16th IFAC symposium on information control problems in manufacturing (INCOM 2018)*.

Espina-Romero, L., Guerrero-Alcedo, J., Goñi Avila, N., Noroño Sánchez, J. G., Gutiérrez Hurtado, H., & Quiñones Li, A. (2023). Industry 5.0: Tracking scientific activity on the most influential industries, associated topics, and future research agenda. *Sustainability, 15*(6), 5554. https://doi.org/10.3390/su15065554

Essakly, A., Wicgmann, M., & Spendgler, T. S. (2019). A reference framework for the holistic evaluation of Industry 4.0 solutions for small and medium-sized enterprises. *IFAC-Papers Online, 52*(13), 427–432.

Erboz, G. (2017). How to define industry 4.0: Main pillars of Industry 4.0. In *7th international conference on management (ICoM 2017)*, At Nitra, Slovakia, November 2017. Retrieved June 9, 2019, from https://www.researchgate.net/publication/326557388_How_To_Define_Industry_40_Main_Pillars_Of_Industry_40

Evans, P. C., & Annunziata, M. (2012). *Industrial Internet: Pushing the boundaries of minds and machines*. Retrieved November 6, 2014, from http://www.ge.com/docs/chapters/Industrial_Internet.pdf

Flores, E., Xu, X., & Lu, Y. (2020). Human cyber-physical systems: A skill-based correlation between humans and machines. In *16th IEEE international conference on automation science and engineering (CASE)* August 20–21, 2020 (pp. 1313–1318).

Fraga-Lamas, P., Lopes, S. I., & Fernández-Caramés, T. M. (2021). Green IoT and edge AI as key technological enablers for a sustainable digital transition towards a smart circular economy: An industry 5.0 use case. *Sensors, 21*(17), 5745.

Frey C. B., & Osborne M. A. (2013). The future of employment: How susceptible are jobs to computerisation? Retrieved November 10, 2023, from http://www.oxfordmartin.ox.ac.uk/downloads/academic/The_Future_of_Employment.pdf

Gajdzik, B. (2023). Industry 5.0 as the upgrade of Industry 4.0: Towards one common concept of industrial transformation. *Organization and Management, Silesian University of Technology, 181*, 131–150. https://doi.org/10.29119/1641-3466.2023.181.9

Gajdzik, B., Siwiec, D., Wolniak, R., & Pacana, A. (2024). Approaching open innovation in customization frameworks for product prototypes with emphasis on quality and life cycle assessment (QLCA). *Journal of Open Innovation: Technology, Market, and Complexity, 10*(2), 1–16. https://doi.org/10.1016/j.joitmc.2024.100268

Gajdzik, B., & Sroka, W. (2012). Analytic study of the capital restructuring process in metallurgical enterprises around the World and in Poland. *Metalurgija, 51*(2), 265–268.

Gajdzik, B., & Štverková, H. (2023). Frameworks of strategic management in Industry 4.0, In *Proceedings of the 15th international conference on strategic management and its support by information systems 2023*, May 22–24, 2023, Ostrava, Czech Republic/Němec Radek, Chytilová Lucie (ed.), 2023, Ostrava, VŠB – Technical University of Ostrava (pp. 72–85) ISBN 978-80-248-4687-3/.

Gajdzik, B., & Wolniak, R. (2022). Influence of Industry 4.0 projects on business operations: literature and empirical pilot studies based on case studies in Poland. *Journal of Open Innovation: Technology, Market, and Complexity, 8*(1), 44. https://doi.org/10.3390/joitmc8010044

Grabowska, S., Saniuk, S., & Gajdzik, B. (2022). Industry 5.0: improving humanization and sustainability of Industry 4.0. *Scientometrics*. https://doi.org/10.1007/s11192-022-04370-1

Gonçalves Machado, C., Winrotha, M. P., & Dener Ribeiro da Silva, E. H. (2020). Sustainable manufacturing in Industry 4.0: An emerging research

agenda. *International Journal of Production Research, 58*(5), 1462–1484. https://doi.org/10.1080/00207543.2019.1652777
Green economy. UN Environment programme. Retrieved from https://www.unep.org/regions/asia-and-pacific/regional-initiatives/supporting-resource-efficiency/green-economy
Groombridge, D. (Gartner analyst). Top strategic technology trends 2023. Retrieved from https://enterprise.press/wp-content/uploads/2022/10/Gartner.pdf
Hauke, K. (2017). Big Data w innowacyjności organizacji biznesowej. Innowacyjność to cyfryzacja i rozwój zarządzanie operacyjne w teorii i praktyce organizacji biznesowych, publicznych i pozarządowych. Prace Naukowe WSZiP, 43(4). Retrieved from http://www.pracenaukowe.wwszip.pl/prace/prace-naukowe-43.pdf
Heuser, L., Nochta, Z., & Trunk, N. C. (2008). *ICT shaping the world: A scientific view*. ETSI, Wiley Publication.
Hermann, M., Pentek T., & Otto, B. (2015). Design principles for Industrie 4.0 scenarios. A literature review. Working Paper No. 01. Technische Universität Dortmund Fakultät Maschinenbau. Retrieved from http://www.iim.mb.tu-dortmund.de/cms/de/forschung/Arbeitsberichte/DesignPrinciples-for-Industrie-4_0-Scenarios.pdf
Hermann, M., Pentek, T. and Otto, B. (2016). Design Principles for Industrie 4.0 Scenarios. Proceedings of 49th Hawaii International Conference on System Sciences HICSS, Koloa, 5–8 January 2016, 3928–3937. https://doi.org/10.1109/HICSS.2016.488.
IFR—World Robotics 2023 Report. International Federation of Robotics. www.ifr.org
Industrial Internet Consortium. (2013). Fact sheet. Retrieved December 3, 2014, from http://www.iiconsortium.org/docs/IIC_FACT_SHEET.pdf
Industry 5.0: Towards a sustainable, human-centric and resilient European industry (7 Jan. 2021). European Commission, Brussels. Retrieved from https://op.europa.eu/en/publication-detail/-/publication/468a892a-5097-11eb-b59f-01aa75ed71a1/
ISO 8374 en. Robotics — Vocabulary. Retrieved September 20, 2023, from https://cdn.standards.iteh.ai/samples/75539/1bc8409322eb4922bf680e15901852d2/ISO-8373-2021.pdf
Kagermann, H. (2015). Change through digitisation—Value creation in the age of Industry 4.0. In: H. Albach, H. Meffert, A. Pinkwart, & R. Reichwald (Ed.), Management of permanent change, edition 127, Chap. 2 (pp. 23–45). https://doi.org/10.1007/978-3-658-05014-6_2
Kagermann, H., Wahlster, W., Helbig, J. (Eds.). (2011). Recommendations for implementing the strategic initiative Industrie 4.0: Final report of the Industrie 4.0. Working Group: Industrie 4.0: Mit dem Internet der Dinge auf

dem Weg zur 4. industriellen Revolution, VDI-Nachrichten, Acatech-National Academy of Science and Engineering: München, Germany. Retrieved from http://forschungsunion.de/pdf/industrie_4_0_final_report.pdf

Kagermann, H., & Winter, J. (2018). The second wave of digitisation (How the internet of things, Industrie 4.0, autonomous systems, and digital business models change the world). In S. Mair, D. Messner, & L. Deyer (Eds.), *Germany and the world 2030. What will change* (pp. 201–209). ECON Publishers.

Kaur, M., Jadhav, A., & Akter, F. (2022). Resource selection from edge-cloud for IIoT and blockchain-based applications in Industry 4.0/5.0. *Security and Communication Networks, 2022*, 9314052.

Klimchik, A., Ambiehl, A., Garnier, S., Furet, B., & Pashkevich, A. (2017). Efficiency evaluation of robots in machining applications using industrial performance measure. *Robotics and Computer-Integrated Manufacturing, 48*, 12–29. https://doi.org/10.1016/j.rcim.2016.12.005

Kowalczyk, L. (2017). Cyfryzacja w procesie postępu cywilizacyjnego i jej współczesna rola w innowacyjności, in: Innowacyjność to cyfryzacja i rozwój zarządzanie operacyjne w teorii i praktyce organizacji biznesowych, publicznych i pozarządowych. Prace Naukowe WSZiP 43(4). Retrieved November 20, 2023, from http://www.pracenaukowe.wwszip.pl/prace/prace-naukowe-43.pdf

Leng, J., Chen, Z., Huang, Z., Zhu, X., Su, H., Lin, Z., & Zhang, D. (2022). Secure blockchain middleware for decentralized IIoT towards Industry 5.0: A review of architecture, enablers, challenges, and directions. *Machines, 10*(10), 858.

Łapuńka, I., & Marek-Kołodziej, K. (2017). Zarządzanie 3.0 w Przemyśle 4.0 (Title in English: Management 3.0 in Industry 4.0). Przedsiębiorczość i Zarządzanie (Entrepreneurship and Management), 18 (10). *18, (10), Part I*, 23–35.

Majerník, M., Daneshjo, N., Malega, P., Drábik, P., & Barilová, B. (2022). Sustainable Development of the Intelligent Industry from Industry 4.0 to Industry 5.0. *Advances in Science and Technology Research Journal, 16*(2), 12–18.

Mayer-Schonberger, V., & Cukier, K. (2014). Big data Rewolucja, która zmieni nasze myślenie pracę i życie. MT Biznes, Warsaw.

McGarry, L., Butterfield, J., & Murphy, A. (2022). Assessment of ISO standardisation to identify an industrial robot's base frame. *Robotics and Computer-Integrated Manufacturing, 74*, 102275.

Miyake, D. (2022). Strategic planning vs. operational planning: 7 main differences. Retrieved January 23, 2023, from https://www.clearpointstrategy.com/strategic-planning-vs-operational-planning/

Nahavandi, S. (2019). Industry 5.0—a human-centric solution. *Sustainability (Switzerland), 11*(16), 4371.
Nakanishi H. (2019). Modern Society has reached its limits - "Society 5.0" Will Liberate us. Davos: World Economic Forum.
Özdemir, V., & Hekim, N. (2018). Birth of Industry 5.0: Making sense of big data with artificial intelligence, "the Internet of Things" and Next-Generation Technology Policy. *OMICS A Journal of Integrative Biology, 22*(1), 65–76. https://doi.org/10.1089/omi.2017.0194
Paprocki, W. (2016). Koncepcja Przemysł 4.0 i jej zastosowanie w warunkach gospodarki cyfrowej. In J. Gajewski, W. Paprocki, J. Pieriegud (Eds.), Cyfryzacja gospodarki i społeczeństwa szanse i wyzwania dla sektorów infrastrukturalnych. Publication of the European Financial Congress, Gdańsk (pp. 39–58).
Paris Agreement: Adoption of the Paris Agreement. United Nations. In *Conference of the Parties Twenty-first session* Paris, November 30–December 11, 2015. https://unfccc.int/resource/docs/2015/cop21/eng/l09r01.pdf
Pahalson, C. A. D., Zwalnan, S. J., & Nimyel, N. C. (2021). Digitisation: Internet of Things, data, service and sustainable strategies in innovation research and technology. *International Journal of Advanced Research in Science, Communication and Technology (IJARSCT), 8*, 1. https://doi.org/10.48175/IJARSCT-1811
Peraković, D., Periša, M., & Sente, R. E., et al. (2019). Information and communication technologies within Industry 4.0 concept. In V. Ivanov (Ed.), *Advances in design, simulation and manufacturing*. Springer International Publishing.
Petrov, K. (2020). Whitepaper: How smart digitisation will help you thrive, Onyx. Retrieved November 20, 2023, from https://onyxinsight.com/whitepapers/how-smart-digitisation-will-help-you-thrive/
Piątek, Z. (2018). Industry 5.0? Na kolejną rewolucję jest raczej zbyt wcześnie. Automatyka, 23 Feb. https://automatykab2b.pl/magazyn
Pieriegud, J. (2016). Cyfryzacja gospodarki i społeczeństwa—wymiar globalny, europejski i krajowy. In: Cyfryzacja gospodarki i społeczeństwa szanse i wyzwania dla sektorów infrastrukturalnych, (red) J. Gajewski, W. Paprocki i J. Pieriegud. Publikacja Europejskiego Kongresu Finansowego, Gdańsk, pp.11–38.
Plattform Industrie 4.0. (2014). Whitepaper FuEThemen. Retrieved November 30, 2014, from http://www.plattformi40.de/sites/default/files/Whitepaper_Forschung%20 Stand%203.%20April%202014_0.pdf
Romero D., Bernus P., Noran O., Stahre J., & Fast-Berglund Å. (2016a). The Operator 4.0: Human cyber-physical systems & adaptive automation towards human-automation symbiosis work systems. In *IFIP international conference*

on advances in production management systems (APMS), Iguassu Falls, Brazil (pp. 677–686).

Romero, D., Stahre, J., Wuest, T., Noran, O., Bernus, P., Fast-Berglund, Å., & Gorecky D. (2016b). Towards an operator 4.0 typology: A human-centric perspective on the fourth industrial revolution technologies. In *Proceedings of the international conference on computers and industrial engineering (CIE46) proceedings*. Tianjin, China.

Rüßmann, M., Lorenz, M., Gerbert, P., Waldner, M., Justus, J., Engel, P., & Harnisch, M. (2015). *Industry 4.0*. The Boston Consulting Group.

Saniuk, S., Grabowska, S., & Gajdzik, B. (2020). Social expectations and market changes in the context of developing the Industry 4.0 concept. *Sustainability, 12*, 1362. https://doi.org/10.3390/su12041362

Senn, C. (2019). The nine pillars of Industry 4.0. Retrieved from https://www.idashboards.com/blog/2019/07/31/the-pillars-of-industry-4-0/

Shala, B., Trick, U., Lehmann, A., Ghita, B., & Shiaeles, S. (2019). Novel trust consensus protocol and blockchain-based trust evaluation system for M2M application services. *Internet of Things*. https://doi.org/10.1016/j.iot.2019.100058

Schwab, K. (2016). The Fourth Industrial Revolution. Publisher: World Economic Forum, Cologny, Switzerland, Polish translation: Kamińska A.D. (2018). Czwarta rewolucja przemysłowa. Publikacja powstała przy współpracy firmy Deloitte. Publishing House: Studio Emka, Warsaw.

Singh, S. K., Yang, L. T., & Park, J. H. (2023). Fusion fed block: Fusion of blockchain and federated learning to preserve privacy in Industry 5.0. *Information Fusion, 90*, 233–240.

Sniderman, B., Mahto, M., & Cotteleer, M. J. (2016). Industry 4.0 and manufacturing ecosystems exploring the world of connected enterprises. Retrieved December 20, 2023, from https://www2.deloitte.com/content/dam/insights/us/articles/manufacturing-ecosystems-exploring-world-connected-enterprises/DUP_2898_Industry4.0ManufacturingEcosystems.pdf

Sriman, B., Annie Silviya, S. H., Santhosh Kumar, E., Suryaa Narayanan, K., & Nishaalu, S. (2022). Blockchain Industry 5.0: Next generation smart contract and decentralized application platform. In *International conference on innovative computing, intelligent communication and smart electrical systems, ICSES.*

Sultan, S., Acharya, Y., & Hynes, N. (2021). Commentary: When promising innovators lack strategic vision: The concept of flow modulation in aortic dissection. *Journal of Endovascular Therapy, 28*(1), 32–35.

Szpringer, W. (2021). Smart industries 5.0: Challenges for regulation (Book Chapter). In *The economics of sustainable transformation* (pp. 193–219).

Taleb, N. (2007). *The Black Swan: The impact of highly impropable*. Random House.

Taleb, N. (2012). *Antifragile: Things that gain from disorder*. Random House.
Tapscott, D. (1999). The digital economy: Promise and peril in the age of networked intelligence. Translated from English by Dubinsky, I. and Pisarev, S. (ed.). Kyiv, ITN Press.
Vor dem Esche, J., & Hennig-Thurau, T. (2014). German digitisation consumer report, digitisation think: Lab Research Report No. 2, Digitisation Think: Lab by Marketing Center Münster and Roland Berger Strategy Consultants, Münster, p. 10. Retrieved December 10, 2023, from https://www.rolandberger.de/media/pdf/Roland_Berger_German_Digitisation_Consumer_Report_20140718.pdf
Yang, H., Asheralieva, A., Zhang, J., Karim, M. M., Niyato D. T., & Raza K. A. (2022). User-centric blockchain for Industry 5.0 applications. In *IEEE international conference on communications workshops, ICC workshops* (pp. 25–30).
Wiesmüller, M. (2014). Industrie 4.0: Surfing the wave? E&I Elektrotechnik und Informationstechnik, 1.
World Robotics Report 2023 Shows Ongoing Global Growth in Installations, Finds IFR. By Robotics 24/7 Staff, September 26, 2023. Retrieved November 20, 2023, from https://www.robotics247.com/article/world_robotics_report_2023_shows_ongoing_global_growth_installations_finds_ifr
Zhu, D., Feng, X., Xu, X., Yang, Z., Li, W., Yan, S., & Ding, H. (2020). Robotic grinding of complex components: A step towards efficient and intelligent machining—Challenges, solutions, and applications. *Robotics and Computer-Integrated Manufacturing, 65*, 101908. https://doi.org/10.1016/j.rcim.2019.101908

CHAPTER 2

Steel Industry in the Era of Digitalisation and Climate Change

Abstract Digitisation and smart technologies, according to prominent researchers, are radically affecting economies, industries and societies all over the world. The digital transformation, of the fourth industrial revolution, dubbed the "second wave" in Industry 4.0 (Kagermann et al., 2018). Digitisation is recognised as the basis for cyber-physical manufacturing systems. Digitisation (i.e. the process of transforming analogue data into digital data sets), combined with technologies attributed to the concept of Industry 4.0 (cloud computing, smart sensors, big data, 3D printing, artificial intelligence to support machine learning, computer-based vertical and horizontal business integration systems, virtual and augmented reality solutions), has opened up new opportunities to bring about changes in processes within companies and across supply chains. Chapter two outlines the impact of digital and smart changes on the steel industry. However, the steel industry's transformation to smart steel manufacturing will require consideration of the requirements of global climate policy (Strategy 2050 "Net Zero"). The global steel industry generates about 7% of global greenhouse gas emissions. To meet its emission targets, the steel industry will need to be climate-neutral by the end of 2050. The deep decarbonisation of the steel industry will result in the replacement of existing high-carbon technologies with low carbon and, in the future, zero-carbon, or green steel.

Keywords Digitisation · Steel industry · Green steel · Decarbonisation

B. Gajdzik, *Digital Transformation towards Smart Steel Manufacturing*,
https://doi.org/10.1007/978-3-031-77756-1_2

2.1 Digitisation of Processes in Steel Industry

Development of the steel industry is supported by ongoing process digitisation and transformation of the steelworks towards Industry 4.0 (see: Neef et al., 2018). Digitalisation offers new opportunities for smart steel manufacturing. Augmented and virtual reality is introduced into processes. Software based on artificial intelligence and machine learning gathers data from the plant and then converts it into information. Digital smelter, as a combination of production processes with information technologies, is able to make better use of opportunities such as: real time, interoperability and horizontal and vertical integration (Gajdzik & Wolniak, 2021a, 2021b; Miśkiewicz & Wolniak, 2020, Peters 2016, 2017; Cheng & Westman, 2020; Santos et al, 2017).

Digital solutions in the steel industry belong to several categories:

- steelmaking technologies (EAF, BOF) and other technologies,
- process handling and work organisation (integrated computer systems),
- product enrichment (product for steel consumers),
- linking physical facilities with digital support production systems (transformation to smart steel manufacturing),
- integration of processes in both steel mills and steel supply chains (vertical and horizontal integration).

Digital technologies in steel mills are used in the following systems (Herzog, 2017; Herzog et al., 2017):

- production control and management systems,
- process optimisation and work organisation systems,
- intelligence maintenance management systems.

Table 2.1 presents the main levels of used computer systems in steel mills.

In industrial transformation, the digitalisation is supported by smart technologies. The support is important to achieve maximum efficiency of both equipment and processes by maximum installation availability, low energy consumption per tonne of steel, minimum scrap quantities and continuous cost-optimised production planning to ensure complete and on-time delivery are key factors for competitive production. Steel

Table 2.1 Levels of used computer systems in steel mills

Level	*Description*
Production control and management system	The PT (Primetals Technology) system is based on the PSImetals product—a set of modules developed specifically for the metal industry. The system is already proven in many metallurgical installations
Process optimisation and quality control system	Through-Process Optimisation (TPO)—an integrated know-how-based concept developed by Primetals Technologies to improve productivity and quality at all steelmaker sites. The digital platform of TPO is the Through-Process Quality Control (abbreviated: TPQC) system, which stores the expertise in quality principles that are applied to all collected high-resolution process data recorded throughout the production chain
Maintenance management system	The Condition Monitoring System (abbreviation: CMS) obtains real-time condition information from the underlying components via traditional sensors or via smart sensors. Condition information can be retrieved offline using a mobile device (e.g. vibration measurements) and fed into the CMS. As a result, the current state of the computer is calculated using the downloaded information. Planned production driving upcoming machine use and the type of products planned for production are considered when predicting the future state of the machine. The current state of the machine and the predicted future state will be made available to the plant's Computerised Maintenance Management System (abbreviated CMMS) as the basis for optimised holistic planning of maintenance activities and maintenance shutdowns. The maintenance plan and production plan are dynamically harmonised between the CMMS and the Predictive Maintenance System (abbreviation: PMS)

Source: Own elaboration based on Herzog et al. (2017), Astor (2016), see Gajdzik (2022a, 2022b)

products are improved through digital and networked processes, additive manufacturing, as well as sustainable and innovative technologies. Steel mills that generate value for steel consumers and the environment can survive in the global steel market. According to McKinsey&Company (Mori et al., 2018) digitalisation of steel manufacturing effect on: energy intensity, throughput, value in use, quality, maintenance, labour productivity, sales, supply chain, procurement and general and administrative (G&A). McKinsey Company in its report fixed the levels of digitisation for steel processes (Table 2.2).

Smart Manufacturing (SM) (with Smart Product (SP) realised in Smart Factory (SF)) is achieved by combining manufacturing systems, analytical systems, network of smart sensors and interfaces, mobile devices with automation and robotisation of activities Cyber-Physical Steel Production System (CPSPS). The framework for smart steel production sets out (Gajdzik, 2022a, 2022b):

- Artificial Intelligence (AI) with Learning Machine (LM) e.g. optical measurement system in additive steel processes (Metal Additive Manufacturing, MAM),
- Additive Manufacturing (AM) based on advanced manufacturing technologies with 3D modeling of products and entire production lines, process tracking system with VR techniques, 3D printing,

Table 2.2 Levels of digital influence on steel processes

Process	*Using*	*Steel manufacturing*	*Rolling and others*
Manufacturing	energy intensity	middle	middle
	throughput	middle	middle
	value in use	high	low
	quality	low	middle
	maintenance	middle	middle
	labour productivity	low	low
Trade and sale	sales	high	high
	supply chain	middle	middle
Procurement	customer services	middle	middle
General and Administrative (G&A)	management	high	high
Total		**high**	**high**

Source: Based on McKinsey Company report (Mori et al., 2018)

advanced metal processing technologies, e.g. Selective Laser Melting (SLM),

- process support IT systems, including ERP (Enterprise Resource Planning) and MES (Manufacturing Execution System) classes with applications tailored to the needs of manufacturing companies, such as ASP-ERP supports purchasing analysis, in cooperation with Industrial Internet of Things (IIoT) applied to industrial automation, equipment integration, inter-device communication, data flow and analytics in the steel industry,
- Big Data and Big Data Analytics with cloud services (Cloud Computing—CC) through access to the Industrial Internet of Things (IIoT) and thematic platforms (operational platforms for data exchange within the plant and across supply chains), as well as Autonomous Control System (ACS) and Automated Decision Making (ADM) metallurgical process control systems,
- systems of operation monitoring based on a network of intelligent sensors and RFID tags to ensure the continuity of equipment operation in accordance with the concept of predictive maintenance of the technology of steel production and processing of steel products (Predictive Maintenance, PM),
- real-time tracking systems for optimizing manufacturing processes with mapping and visualisation for physical objects, including virtual reality (VR) based on the virtual twin of the physical world (VT) for testing production lines and new products (prototyping),
- digitisation of the supply chain with blockchain technology and Life Cycle Management (LCM),
- digital business networks with access to Big Data and IIoT, as well as integrated information and communications technologies (ICTs) and intelligent process tracking systems in steel chains.

In digitalisation, steelworks focus on interoperability and process optimisation. In the implementation of the computer systems, steelworks cooperate with IT companies that help them digitise processes and show the possibilities of digital technologies. Data from machines, installations, facilities and products are collected in the MES (Manufacturing Executing Systems) system. MES, together with ERP, creates a vertical system of process connections and data from the devices is sent to process analysis systems. In the digitalisation of processes, the following services are also useful: Google Cloud, Microsoft Azure. The IT systems used are constantly expanded and improved. All used computer systems create an integrated management system (Gajdzik & Wolniak, 2021a, Torn & Vaneke, 2019, Gajdzik, 2022a, 2022b).

Digitisation in smelters (steel mills) is aimed at comprehensive support of all processes, from material management and stock control through synchronisation of deliveries with production planning, to control of manufacturing and transport of steel products to customers. Steel manufacturing, from the blast furnace to the rolling mill, is computer controlled (manual or autonomous). The key system of process support is the enterprise resource planning (ERP). The system is used, among other things, to improve the efficiency of the management of raw materials and energy intensity according to the 3R principle (Reduce, Recycle, Reuse). Optimising processes by saving resources is an important part of the sustainable steel production. During steel production, steel mills use the following computer systems: optical inspection of diameters of rolled profiles of steel products in the online system; measurement the diameters of metal components and the width of metal plates by LDM sensors; detection and control of the position of hot products by an infrared product scanning system (Rota-Sonde); detection of defects in the product structure by means of triangulation scanning (Trilas TL). Moreover, logistics processes in steel manufacturing are controlled by RFID and other solutions, including Dials FT (this chip studies the flight time and provides data on distances and dimensions between steel products). Additive manufacturing is carried out on 3D printing and precision laser and welding equipment. Stereoscopic imaging (DigiScan) together with a width measurement and image processing (CropVision) is used in steel mills for cutting steel products. The novel solutions are more precise in performing machining operations than the technological solutions used so far. State-of-the-art production equipment with a wide range of wireless controllers and sensors are more and more autonomous. The digitisation of metallurgical processes is realised at all business levels in the enterprise. There is no one universal solution that covers all areas of digitisation applications in the steel plant. Changes are phased in according to the importance of the processes, from key to additional ones. The final form of changes will be a new system architecture called the cyber-physical steelmaking system (CPSPS) (more in Iannino et al., 2019).

2.2 Background of Smart Steel Manufacturing

Intelligent or smart manufacturing, defined as the integration of manufacturing with modern information technologies such as 5G, digitalisation, networking and intelligence, has grown in popularity as a means of boosting the productivity, intelligence and flexibility of traditional manufacturing processes (Zhou et al, 2022). According to Kusiak (2017), "smart manufacturing integrates manufacturing assets of today and tomorrow with sensors, computing platforms, communication technology, data intensive modelling, control, simulation and predictive engineering". "Smart manufacturing utilises the concepts of the cyber-physical systems, Internet of Things (and Everything), cloud computing, service-oriented computing, artificial intelligence and data science" (Kusiak et al., 2017, p. 509).

The main implementers of smart manufacturing are key technologies of Industry 4.0 (Big Data, cloud computing, IoT, AI) and human pace in the cyber-physical systems (Wang et al., 2021). In the smart manufacturing, the machines are connected to each other, but also to the Internet to ensure optimal and scalable control of production processes. The aim of the smart manufacturing is to create new value for the customers by opportunities of new manufacturing technologies.

The steel industry is an indispensable support for modern life and economic development, with global steel industry output approaching 2 billion tonnes of steel. According to the World Steel Association (report: World Steel in Figures, 2023), global crude steel production reached 1885 million tonnes in 2022. The global steel industry employs 6 million people (WorldSteel: https://www.worldsteel.org/). The European steel industry directly employs around 326,000 people, almost 5% of the total workforce in the global steel sector. Additionally, the European steel industry creates employment for 2.67 million people working in related industries (European Steel in Figures, 2023). The European steel industry could produce 170 million tonnes of steel per year, but production has declined in recent years. According to the World Steel Association, steel production in EU countries in 2022 fell by 10.5% year-on-year—to 136.7 million tonnes (report: World Steel in Figures, 2023).

Governments around the world have included smart manufacturing in their national development plans, with the aim of increasing the level of intelligence in the manufacturing industry, creating a smart factory with personalised customisation and optimal resource allocation that adapts to large specifications and integrating customers and business partners into business and value processes to increase productivity. The concept of smart manufacturing enters to steel mills, which want to more competitive in the global steel market.

Yin (2017) proposed that the smart steel manufacturing should be based on the physical input and output of the geophysical system's material, energy and information flow networks to optimise the physical system structure of the manufacturing process and integrate the digital information system to achieve the intelligence state. Yao and Wang (2020), Liu (2020), Li and Li (2021) suggested a steel intelligent factory composition analysis, steel process Big Data, steel production value chain restructuring and artificial intelligence to boost intelligent manufacturing. Other authors, e.g. Govender (2019), Shin and Park (2019) and Zheng et al. (2022) described the approach for implementing an Industry 4.0 framework and a human–cyber-physical (H-CPS) system for production and operation decision optimisation in the steel industry.

The pathway of steel mills towards smart steel manufacturing is an individual decision for each company. Companies decide what process will be changed and what investments will be realised (Gajdzik et al., 2021). Steel mills employ, especially during the first pilot projects, digital companies, which help them to build the smart steel manufacturing. Lots of organisations provide off-the-shelf digital solutions and packages of smart technologies and services. The specialised companies help manufacturers to build cyber-physical production systems (CPS). Digital companies help manufacturers to choose proper technologies and implement them. In steel production can align several layers smart manufacturing: (1) device, (2) network, (3) Big Data, (4) application. These layers were pressed by Zhou et al. (2022) based on Chine steel industry (Table 2.3).

According to World Steel "smart manufacturing does not just mean having a steel-producing smart factory. It is a significant transformation in the way we source raw materials, manufacture and market our products through horizontal and vertical supply chain integration—and it is profoundly customer-focused". Levels of vertical integration in the steel industry are presented in Table 2.4.

Changing market mechanisms is a strong driving force which will require flexibility of operations and mass-customisation of products. Moreover, some steel products will be produced in small batch-size and short lead-time. Surfing the waves of this change will require the steel industry to have a sharp awareness of customer dynamics and expertise

Table 2.3 Layers of smart steel manufacturing

Layer	*Description of solutions*
Device	Systems for analysing raw material, process and quality parameters. Equipment analysis systems: condition monitoring, temperature detection, defect detection, online composition control, shape detection and other critical process parameters should be detected online In this way, DL can provide a huge amount of real-time data for production control. Data is a solid foundation for big data and smart manufacturing applications
Network	Due to the steel process's big-scale, multi-equipment and prolonged process features, the physical distance between each piece of equipment is long, and huge data in diverse formats must be transmitted in real time across a significant number of networks. Data transport is mostly accomplished using optical fiber transmission, industrial network (Wi-Fi, 5G)
Big Data	Big data platform for steel manufacturing processes with process modelling, Big Data analytics, data management, data visualisation
Application	Visualisation of production status, intelligent logistics tracking, equipment status monitoring, order digitalisation and intelligent scheduling, digital process simulation optimisation, intelligent decision-making, online monitoring the control of the whole process, energy management and optimisation

Source: Own elaboration based on Zhou et al. (2022)

in the implementation of tools and strategies offered through Industry 4.0. Smart manufacturing is a transition from a production organisation to a steel consumer-oriented manufacturing organisation. Such solution as a virtual simulation system for the entire ironmaking helps consumers in the improvement of steel products. Steel consumers are active participants in the manufacturing process by using the digital technologies and computer communication systems (more in: Gajdzik et al., 2025).

2.3 Towards Smart Steel Supply Chain

Supply chains (SC) play a crucial role in manufacturing and service operations. "Supply chain management (SCM), which is a systematic approach to managing flows of assets from sourcing raw materials, product manufacturing, to delivering to end customers significantly affects the business goals of the partners in supply networks "(Zhang et al., 2023).

The development of supply chains in the steel industry is currently influenced by changes in the global market. In the global steel market,

Table 2.4 Solutions of vertical integration in steel industry

Level of integration	*Description of changes*
Logistics	• real-time tracking of supplies and orders assisted through GPS, RFID, Wi-Fi (indoor & outdoor accurate positioning) • dynamic management of in-process inventory, consumables and spares
Product quality systems	• modelling of material and surface characteristics for local process control resulting in optimum quality • Computer Aided Quality Control (CAQC) with 100% testing, inspection and storage of data related to hot-rolled coils minimising claims and cost of potential quality issues • intelligent grading of coils and rerouting of coils within production lines to find best match with customer orders reducing rejects and optimising yield • operational research (Big Data analytics) of process data related to coil segments for fault tracking and process efficiency • real-time dynamic line scheduling and adjustment of process parameters with Artificial Intelligence (BIOMIMIC + Parallel Computing) in rolling and finishing
Predictive maintenance	• predictive asset maintenance increasing up-time assisted by advance warnings of failures • remote assistance of maintenance teams with smart glasses
Process control and safety	• dynamic real-time analysis and control of blast furnace process parameters using sonic, laser and radar visualisation • sensors detect harmful stray gases, noises and temperatures, and notify operators of threats • advanced fully automated BOF from charging to tapping • networked platforms of assets such as blast furnaces where process experts can see every operation in real time and collaborate with each other • use of drone technology for inspection of areas that are difficult to reach in the plant and surveying and planning of mining operations

Sources: Examples from World Steel Association. Retrieved from https://worldsteel.org/about-us/smart-manufacturing

there are more steel products and services. Moreover, the development of this market is strongly influenced by information technologies and the global policy of industrial decarbonisation. Since the adoption of the Paris Agreement, there has been a substantial growth in climate-related targets in the global steel sector. Moreover, national-level pledges announced to date to achieve net zero GHG or CO_2 emissions by 2050–2070 cover the entire global steel production (Net Zero Tracker, 2022; World Steel Association, 2020).

The business environment is dynamic and even periodically unstable (COVID-19 pandemic). In such an environment, steel supply chains need to be increasingly flexible and intelligent. Resilience is identified as a strategic direction for Industry 5.0 (see: European Document: Industry 5.0: Towards more sustainable, resilient, and human-centric industry, 7 Jan. 2021). Using Industry 4.0 technologies, traditional supply chains are becoming increasingly smart. More objects equipped with sensors and ICT are appearing in the supply chain. The smart supply chain needs information management, IT infrastructure, process automation, advanced analytics and supply chain integration (Wu et al, 2016). Intelligent decision-making processes are structuring new smart supply chains. IT systems based on a network of collaborative applications located in individual nodes of the chain, control and operate business processes extend along the entire chain and are not limited to one organisation (Rutkowski, 2004, p. 59). Innovative technologies offer unprecedented opportunities to reduce costs and improve efficiency. A smart supply chain makes optimal use of the resources of the actors in the chain and responds quickly to changes in the environment. This response to change is made possible by access to real-time data (Barcik and Odlanicka-Poczobutt, 2020). Every smart supply chain (SSC) needs advanced technologies, especially emerging ICT, to link the processes in the different partners of a SC to form an intelligent connected system (Wu et al., 2016).

Collaborative structures in the steel industry are a constellation of multiple business entities. Business processes are supported by digital technologies and communication, analysis and control systems. Smart technologies make it easier for entrepreneurs to build a common operating mechanism based on multiple configurations of business relationships. The digitization of the chain, due to the wide range of links between actors, should be geared towards the compatibility of IT solutions (Gajdzik et al., 2024; Gajdzik, 2022a). The main challenge in the digitisation of the steel industry is the integration of all IT and production support systems. Integration takes place in three dimensions (levels): (1) enterprise level: vertical integration supported by production automation, autonomous sensors and ERP systems, (2) chain level: horizontal integration based on integration of systems along entire chains, (3) life cycle level (life cycle integration): integration across the entire life cycle product from manufacturing engineering to waste management and recycling (Miśkiewicz & Wolniak, 2020).

The ongoing digitalisation of processes in steelmaking chains is often the result of pressure from powerful capital groups. Steel mill owners have created strong cooperative environments. The combined capital in the steel industry has created global organisational structures. Steel chains are formed by specialised production companies, supply and delivery companies, distribution centres and entities, coking plants, power plants, etc. Chains create companies belonging to common capital group and other companies based on commercial agreements (Gajdzik, 2016; Gajdzik & Sroka, 2012). Capital- and business-related companies in the steel market seek process flexibility throughout the supply chain through distributed-type cooperation networks. Network participants in the steel industry benefit from global communication networks and advanced data exchange technologies in cyberspace. According to Brettel et al. (2014) visualisation, networking and decentralisation are creating a new business space.

The changed supply chains according to the idea of Industry 4.0 are called supply chain 4.0. Mc Kinsey Company (2016) has proposed the following definition: "Supply Chain 4.0 - the application of the Internet of Things, the use of advanced robotics, and the application of advanced analytics of big data in supply chain management: place sensors in everything, create networks everywhere, automate anything, and analyse everything to significantly improve performance and customer satisfaction". A new supply chain—Supply Chain 4.0—will be faster, more flexible, more personalised, more accurate, more efficient (Table 2.5). Such smart supply chain involves automated systems including Artificial Intelligence (AI), robots, drones, nanotechnologies and a variety of inputs that enable customization, flexibility and rapid manufacturing, advanced automation in warehouses and new forms of transport. Logistics is taking a huge step forward with better connectivity, advanced analytics, additive manufacturing and advanced automation. Autonomous and intelligent vehicles are on the rise in automated warehouses, and customers are participating in product design and process tracking.

Modern supply chains are constantly being adapted to global trends and legislative requirements. In the era of Industry 4.0/5.0, chain companies are investing in technologies that change the way they manage and optimise processes. The transformation into a digital and smart supply chain requires two key enablers—technologies and environment. This

Table 2.5 Results of used new technologies in supply chains

Results	*Solutions*
Faster	• advanced forecasting approaches, e.g. predictive analytics of internal (e.g. demand) and external (e.g. market), • shipment tracking system (delivery), • blockchain technology, • autonomous and smart vehicles (self-guided vehicles), • smart automation, • lifecycle system
More flexible	• real-time production capacity feedback from machines, • continuous process planning, • advanced analytics, • shipment tracking system (delivery): redirecting shipments to the most convenient destination, • predictive analytics, • flexible transport, • blockchain technology, • visualisation, • predictive maintenance and 3-D printing, • advanced robotics solutions, • smart warehouses
More personalised	• system communication (24 h), • linked (compatible) information computer systems, • prototyping and digital twin, • interoperable technologies
More accurate	• performance management systems provide real-time, end-to-end transparency throughout the supply chain (synthesised top-level KPIs), • additive manufacturing, • autonomous operations systems, • advanced robotics solutions, • integration of data of suppliers, service providers in a supply chain cloud, • performance management systems that "learn" to automatically identify risks or exceptions and will change supply chain parameters in a closed-loop learning approach to mitigate them
More efficient	• automation of both physical tasks and planning, • autonomous robots (robots manage the material (pallets/boxes as well as single pieces) completely automatically along the warehouse process—from receiving/unloading to putting away to pick, pack and ship), • autonomous trucks transport the products within the network, • • transport optimisation systems with the function of sharing capacities between companies, • • dynamic planning approaches to drive advanced demand shaping activities (e.g. unique offers for delivery time slots with low truck utilisation)

Source: Own elaboration

supply chain needs to provide a high degree of flexibility. The features of Supply Chain 4.0 should be fast, flexible and efficient. These features need state-of-the-art IT systems (Gajdzik & Wolniak, 2024).

The supply chain in steel sector is built up of fixed components, which are suppliers of materials: iron ore, scrap, energy, etc., key customers (steel consumer markets), key steel producers (steel mills with steel melting technologies), producers of steel products and steel distribution companies. Steel products in value chains are strongly differentiated. The differentiation thresholds of metallurgical products apply to basic and special-purpose products. R&D departments and research laboratories design products for specific customer orders. Manufacturers of steel products, considering the needs of markets, differentiate many end operations of the production cycle, such as a wide range of products with different dimensions, profiles, colours, etc. Packaging and product labelling are also being differentiated to suit the requirements of the customer market. Offering, by steel mills, highly differentiated products require the design of a comprehensive material flow system both in the production enterprise and throughout the supply chain (integrated ERP management systems are facilitating steel mills at this stage).

The challenge for the steel sector is blockchain technologies (BTs), as blockchain data on steel products in supply chains. Companies (suppliers, manufacturers, distributors, consumers), collaborate in the supply chain to source raw materials, convert them into finished products, or deliver products to consumers (Ayan et al., 2022). Supply chains are getting longer and more complicated as they become more global. One of the challenges that modern supply chains typically face is sustainability, and BT is seen as essential to a company's ability to achieve supply chain sustainability (Ayan et al., 2022). BT was first conceptualised in 2008 by a person (or group) named Satoshi Nakamoto, in an article titled "Bitcoin: Peer-to-Peer Electronic Cash System" (Crosby et al., 2016, p. 9). Blockchain can be characterised as a decentralised shared database in which verified and synchronised data is stored in a peer-to-peer network using chronological, encrypted, chained blocks. Such blockchain can manage safety, corporate responsibility, environmental protection, manufacturing transparency, full life cycle tracking of steel products (Cao et al., 2020).

2.4 Decarbonisation of Steelmaking Process Towards "Green Steel"

The global steel industry generates about 7% of global greenhouse gas emissions. To meet its emission targets, the steel industry will need to be climate-neutral by the end of 2050 (Climate-neutral Steelmaking in Europe Decarbonisation pathways, Nov. 2021). Achieving this goal will require, on the one hand, improvements in material efficiency and energy savings in the steelmaking process, achieved through advanced IC technologies and smart solutions, and on the other hand, investment in carbon capture storage (CCS) and utilisation (CCU) solutions and cleaner production processes (GREENSTEEL for Europe, Nov. 2021). The project—GREENSTEEL—provides insights and recommendations for effective clean steelmaking solutions suitable for the EU to meet its 2030 climate and energy targets and to implement a long-term strategy for a climate-neutral Europe in 2050. The entire European steel industry is currently striving to reduce direct and indirect CO_2 emissions. The current goal is to reduce CO_2 emissions by 80–95% in 2050 compared to 1990 levels. However, this change cannot be an immediate change. This process is iterative and will require adjustments as it is implemented. The steel industry changes towards climate-neutral technologies. Such a radical technological change is a huge technical investment. The investment is linked to the production, storage, transmission and use of hydrogen. Steel mills must provide their manufacturing processes with green energy. Access to clean energy is needed to reduce emissions by 80–95%. For the viability of planned technological change, it is crucial to ensure access to low-carbon electricity at competitive prices and financial and organisational support from government institutions (Decarbonisation Pathways, Nov. 2021).

In the climate transformation—deep decarbonisation, smelters using coal or gas may have to switch to "green hydrogen" (Decarbonisation Pathways, Nov. 2021). The strategic transformation of the steel industry is to move towards the steel production with green hydrogen as a reducer. Green hydrogen does not emit CO_2 during combustion, unlike traditional fossil fuels—such as coal and natural gas—which are currently used in the steelmaking process. Thus, the use of green hydrogen can help reduce greenhouse gas emissions and reduce the negative environmental impact of industry (Kiessling et al., 2022; Pike, 2017).

Gajdzik et al. (2023a) describe two basic scenarios of deep decarbonisation in steel industry, the first is connected with EAF technology (Electric Arc Furnace)—steel mills will invest in a new generation of EAF furnaces (newly built furnaces must be capable of running on anhydrous hydrogen using renewable energy), the second for BF-BOF technology (Blast Furnace—Basic Oxygen Furnace)—steel mills will replace BF-BOF technology by new one—DRI technology (direct reduction technology). Steel mills must modernise traditional technology—blast furnace with oxygen converter—(BF-BOF) by 2050. In place of this technology will be a new direct iron reduction: DRI. The construction of new plants based on direct reduction of iron ore (DRI process) technology consists of two stages. In the first stage, coke oven gas and natural gas is used. In the second green hydrogen is to function as a reducing agent. The DRI process is to be simultaneously combined with large-scale electric furnaces employing EAF technology. These furnaces will utilise sponge iron feedstock obtained through the DRI process, thus forming the DRI-EAF system. Electric arc furnace technology for which the feedstock is scrap (secondary production) or sponge iron in DRI technology will be continued and developed. Due to its high efficiency and maturity, no revolutionary change is envisaged in this regard, but only continued progressive improvement, particularly in the direction of improving energy efficiency and environmental impact (Gajdzik et al, 2023b).

On the path to deep decarbonisation will be realised the additional changes, the following milestones will be pursued (Podsiadło et al., 2023):

- adaptation of fossil fuel-based steel production through process integration;
- capture and use of waste coal to produce chemicals, e.g. the Steelanol project;
- increasing the recycling of steel scrap and steel by-products.

Technologies in industries can be divided by their impact on the way production is conducted into two groups (Christensen, 1997):

- sustaining and adaptive technologies: these technologies are able to keep some incremental improvements, maintaining the status quo of processes and business scopes;
- disruptive technologies: these technologies can bring significant operational benefits and radically change the ways and business scopes.

In deep decarbonisation of steel industry, the steel mills will have to adopt technologies that will be (are already) classified as subversive, or disruptive technologies. The pace of adaptation of disruptive technologies in the steel industry may gradually accelerate. Companies that understand the essence of technological change too late may lose their current position in the steel market. Breakthrough technologies for decarbonising the steel industry will spawn new manufacturing, as this is the case with many technologies that the world has deemed subversive. The new manufacturing process is called green because it is linked to low-carbon and zero-carbon steelmaking technologies. Steel produced with zero-carbon technologies is called green steel. Green steel will be without carbon footprint.

The term "green steel " is used to steel products manufactured using less GHG-intensive production processes (Muslemani et at, 2021). Green steel is differentiated from "sustainable steel" because the second term—sustainable steel—besides lower GHG emissions also encompasses other issues such as energy and resource efficiency, circularity and reduction of other pollutants (Muslemani et al., 2021).

Green steel can be achieved through various technological pathways (Table 2.6), some of which may be more suitable for specific producers and regions, depending on local factors related to energy infrastructure and demand. EU policy has a significant role to play in the decarbonisation of the steel industry. Nevertheless, member state environmental, energy and industrial policies can also affect the prospects for certain industrial decarbonisation pathways. In the long term, some decarbonisation technologies may end up being more successful and competitive than others. In Table 2.6 are presented the key technologies proposed in the roadmaps to decarbonisation according to the GREENSTEEL project. The nine CO_2 mitigation technologies are proposed for steel industry: (i) hydrogen-based direct reduction, (ii) hydrogen plasma smelting reduction, (iii) alkaline iron electrolysis, (iv) molten oxide electrolysis, (v) carbon oxide conversion, (vi) iron bath reactor smelting reduction, (vii) gas injection into the blast furnace, (viii) substitution of fossil energy carriers by biomass and (ix) high-quality steelmaking with increased scrap usage. Key technological changes in steel manufacturing complemented by auxiliary technologies include CO_2 capture (CCS—Carbon Capture Storage)/ usage (CCU—Carbon Capture and Usage) and hydrogen generation from grey (with CO_2 emissions, through blue: produced from fossil fuels using carbon capture and storage technology, less polluting than grey and turquoise—produced by pyrolysis from natural gas, but it is still a fossil fuel and as such is not emission-free, to green—a clean fuel that allows the storage and use of energy from renewable sources. Finally, four promising technology routes for climate-neutral steelmaking were identified: optimised Blast Furnace—Basic Oxygen Furnace (BF-BOF) (Route 1), direct reduction (Route 2), smelting reduction (Route 3) and iron ore electrolysis (Route 4).

The four key stakeholders in decarbonising the steel industry are steel producers, governments, taxpayers, steel consumers. Their activities are as follows: (portal: zlom-info.pl, 7 Dec. 2021).

- Steel producers must invest in modern technologies that reduce CO2 emissions to produce green steel. Technological innovation is a determinant of being in the global steel market. Steel mill owners should engage private funds for steel mill transformation, as well as EU funds and look for other sources of investment financing.
- Governments must monitor changes and continuously adapt their policies to changing business conditions. They should also provide support through access to knowledge-sharing platforms and invite investors to co-create knowledge about industry transformation. The key elements of their policies should be economic, financial, informational, cooperative instruments.
- Taxpayers: The Green Steel for Europe consortium estimates that some €2 billion in EU subsidies will be made available to the steel industry between 2021 and 2030 to help combat climate change and meet emissions targets. EU member states should participate in further financing of the transition. In addition, regulations and tax-funded subsidies are needed to enable the transition.
- Steel products consumers will pay most of the costs of the green transition. Steel consumer markets must already build awareness among end users of their products, e.g. cars, houses, buses, ships,

Table 2.6 Decarbonisation technologies in the Green Steel for Europe (GREENSTEEL)

CCU + CCS		*Carbon direct avoidance*	
Iron bath reactor smelting reduction	Reducing agent—carbon Current TRL—TRL 6 Project examples: HIsarna	Hydrogen-based direct reduction	Reducing agent—hydrogen Current TRL—TRL 6–5 Project examples: Hybrit, Salcos, HYFOR
Gas injection into the blast furnace	Reducing agent—carbon Current TRL—TRL 5–8 Project examples: ULCOS	Hydrogen plasma smelting reduction	Reducing agent—hydrogen Current TRL—TRL 6–5 Project examples: SuSteel
Alternative carbon sources	Reducing agent—carbon Current TRL—TRL 7 Project examples: SHOCOM, GREEN EAF2	Alkaline iron electrolysis	Reducing agent—electricity Current TRL—TRL 5–6 Project examples: SIDERWIN, ASCoPE, IERO
Increased scrap used	Reducing agent—(carbon) Current TRL—TRL 4–8 Project examples: ADAPTEAF, FLEXCHARGE	Molten oxide electrolysis	Reducing agent—electricity Current TRL—TRL 2 Project examples: IERO, VALORCO, ULCOS
Carbon oxide conversion	Reducing agent—carbon Current TRL—TRL 8 conversion, TRL 4–5 (implementation) Project examples: STEELANOL, Carbon2Chem		

TRL Technology Readiness Levels

etc., as well as position their new products in the market by calling them "green" or using other terms for climate-friendly products.

How much stakeholders have to pay for the green transformation will depend on factors such as increasing external pressure to reduce greenhouse gas emissions and the effectiveness of reforms such as the EU's Carbon Boundary Adjustment Mechanism (CBAM), which aims to create a level playing field from 2023 for European industry, mainly by charging non-EU players based on their CO2 emissions and others. The environment in which the business operates has been very labile in recent years (pandemic and war in Ukraine), so one has to look carefully at the designated decarbonisation paths. In Table 2.7 are presented the recent projects on the path of steel industry towards strategy "net zero".

During the deep decarbonisation, the steel industry can achieve the following benefits (Gajdzik et al., 2023b based on literature review and own analysis):

Decarbonising the steel industry can lead to significant reductions in greenhouse gas emissions, particularly carbon dioxide (CO_2), which is one of the main drivers of climate change. By transitioning to low-carbon alternatives, such as hydrogen-based direct reduction or carbon capture and storage, the steel industry can produce green steel.

Decarbonising the steel industry can reduce its dependence on fossil fuels, such as coal and natural gas, which are subject to price volatility and supply chain disruptions. By cooperating with suppliers of renewable energy sources (RESs), the steel industry can increase its energy security.

The transition towards green steel production can create new jobs in areas such as renewable energy, hydrogen production and carbon capture and storage. Moreover, increasing recycling and reuse of steel products can also create new jobs in the circular economy. Steel scrap is the main material for electric furnaces (EAF technology). It can be assumed that with the development of this technology and the phasing out of BF + BOF technology, the steel industry will need more steel scrap, which is why it needs to be recovered more and more.

Besides deep reduction of CO2 emissions, the use of alternative fuels in steel production can reduce emissions of particulate matter, sulphur dioxide and nitrogen oxides, which are harmful to human health.

Table 2.7 Key green steelmaking initiatives

Company, project, place	*Description of technological project*
ArcelorMittal/Hamburg, Germany/Bremen, Germany/Dunkirk, Germany/ Asturias, Spain	Hydrogen reduction with grey hydrogen derived from natural gas Blast furnace + electrolysis for hydrogen production Hybrid blast furnace with direct reduced iron (DRI) gas injection Coke oven gas with grey hydrogen; hydrogen in DRI-EAF
HYBRIT (SSAB, LKAB and Vatenfall) Sweden	Replacing coking coal with hydrogen and fossil-free electricity
Boston Metal, BHP & Vale China Baowu & BHP	Molten oxide electrolysis (MOE) Memorandum of Understanding to share technical expertise in reducing emissions from steel, including using CCUS
Ovako, Sweden	Hydrogen use to heat steel before rolling
Hofors, Liberty Ostrava	Building hybrid furnaces
Rogesa/Germany	Hydrogen in coke gas as reducing agent
Tata Steel/Netherlands	Carbon capture and storage under North Sea; water electrolysis to produce hydrogen and oxygen
Thyssenkrupp/Duisburg, Germany	Hydrogen as reducing agent; use of renewable hydrogen from RWE & feasibility study for water electrolysis plant
Voelstalpine Primetals Tech./Linz, Austria	Hydrogen as reducing agent to process iron ore concentrates
Tenaris, Edison and Snam/Bergamo, Italy	Electrolysis/hydrogen-based steelmaking
Salzgitter AG (Salcos project)/ Wilhelmshaven, Germany	Electrolysis/hydrogen-based steelmaking
Duferco/Brescia, Italy	Using hydrogen in beam furnace with green PPA
Celsa/Statkraft & Mo industrial park AS/ Norway	Electrolysis/hydrogen-based steelmaking
H2 Green Steel Initiative/Northern Sweden	Hydrogen-based steelmaking

Source: Muslemani et al. (2021)

Decarbonising the steel industry can stimulate economic growth by creating new jobs, promoting innovation and attracting investments in innovative technologies and infrastructure.

Low- or zero-emission steelmaking technologies can also lead to increased competitiveness and market access in a rapidly changing global economy. Green steel will become a new product, but the environmental awareness of steel consumers' needs to be built.

References

ASTOR. (2016). Hajkuś J. W jakie technologie inwestują firmy produkcyjne w Polsce?, Biuletyn Automatyki, 87. Retrieved from http://www.astor.com.pl/downloads/biuletyn_automatyki/ Biuletyn_Automatyki_87.pdf (2016-05-16).

Ayan, B., Güner, E., & Son-Turan, S. (2022). Blockchain technology and sustainability in supply chains and a closer look at different industries: A mixed method approach. *Logistics, 6*, 85. https://doi.org/10.3390/logistics6040085

Barcik, R., & Odlanicka-Poczobutt, M. (2020). *Logistyka 40 - wybrane zastosowania*. TNOiK.

Brettel, M., Friederichsen, N., Keller, M., & Rosenberg, M. (2014). How virtualization, decentralization and network building change the manufacturing landscape: An Industry 4.0 perspective. *International Journal of Mechanical, Aerospace, Industrial, Mechatronic and Manufacturing Engineering, 8*(1), 37.

Cao, Y., Jia, F., & Manogaran, G. (2020). Efficient traceability systems of steel products using blockchain-based industrial internet of things. *IEEE Transactions on Industrial Informatics, 16*(9), 6004–6012. https://doi.org/10.1109/TII.2019.2942211

Cheng, J., & Westman, J. (2020). Effects of digitalisation in steel industry economic impacts & investment model. Master of Science Thesis TRITA-ITM-EX 2020:280 KTH Industrial Engineering and Management Industrial Management, Stockholm (pp. 35, 39).

Christensen, C. (1997). *The innovator's dilemma: When new technologies cause great firms to fail*. Harvard Business School Press.

Climate-neutral Steelmaking in Europe Decarbonisation pathways—Investment needs—Policy Conditions—Recommendations, European Commission, Brussel, November 2021, https://op.europa.eu/en/publication-detail/-/publication/89cda1b2-b169-11ec-83e1-01aa75ed71a1/language-en/format-PDF/source-254269180

Crosby, M., Pattanayak, P., Verma, S., & Kalyanaraman, V. (2016). Blockchain technology: Beyond bitcoin. *Applied Innovation, 2*, 71.

European steel in figures 2023. Eurofer. Retrieved November 20, 2023, from https://www.eurofer.eu/statistics

Gajdzik, B. (2022a). Diagnoza kierunków transformacji przemysłu stalowego w Przemyśle 4.0 (Diagnosis of the steel Industry transformation direction in the

Industry 4.0). Monograph 945. Silesian University of Technology, Gliwice 468 p.

Gajdzik, B. (2022b). Frameworks of the maturity model for Industry 4.0 with assessment of maturity levels on the example of the segment of steel enterprises in Poland. *Journal of Open Innovation: Technology, Market, and Complexity, 8*, 77. https://doi.org/10.3390/joitmc8020077

Gajdzik, B. (2016). Sieciowa organizacja grup kapitałowych w krajowym sektorze hutniczym. *Organizacja i Zarządzanie, Kwartalnik Naukowy Politechniki Śląskiej, 1*(33), 5–22.

Gajdzik, B., Grabowska, S., & Saniuk, S. (2021). A theoretical framework for Industry 4.0 and its implementation with selected practical schedules. *Energies, 14*(4), 1–24.

Gajdzik, B., & Sroka, W. (2012). Analytic study of the capital restructuring process in metallurgical enterprises around the World and in Poland. *Metalurgija, 51*(2), 265–268.

Gajdzik, B., Sujová, E., & Biały, W. (2023a). Decarbonisation of the steel industry: Theoretical and practical approaches with analysis of the situation in the steel sector in Poland. *Acta Montanistica Slovaca, 28*(3), 621–636. https://doi.org/10.46544/AMS.v28i3.08

Gajdzik, B., & Wolniak, R. (2021a). Digitalisation and innovation in the steel industry in Poland—Selected tools of ICT in an analysis of statistical data and a case study. *Energies, 14*(3034), 1–25. https://doi.org/10.3390/en1411303410.3390/en14113034

Gajdzik, B., & Wolniak, R. (2024). Changes in the steel chain in Industry 4.0. Some results of survey on the Polish steel market. *Production Engineering Archives*, Sciendo, *30*1, 1–16, https://doi.org/10.30657/pea.2024.30.1

Gajdzik, B., & Wolniak, R. (2021b). Transitioning of steel producers to the steelworks 4.0—Literature review with case studies. *Energies, 14*, 4109. https://doi.org/10.3390/en14144109

Gajdzik, B., Wolniak, W., & Grebski, W. (2023b). Process of transformation to net zero steelmaking: Decarbonisation scenarios based on the analysis of the Polish steel industry. *Energies, 16*(3), 3384. https://doi.org/10.3390/en16083384

Gajdzik, B., Wolniak, R., & Grebski, W. (2024). Changes in the steel chain in Industry 4.0. Some results of survey on the Polish steel market. *Production Engineering Archives, 30*(1), 1–16. https://doi.org/10.30657/pea.2024.30.1

Gajdzik, B., Wolniak, R., Grebski, W., Szymszal, J., & Grebski, M. E. (2025). *Sustainability and smart manufacturing the transformation of the steelwork industry*. Routledge.

Govender, E., Telukdarie, A., & Sishi, M. N. (2019). Approach for implementing industry 4.0 framework in the steel industry. In *Proceedings of the 2019 IEEE international conference on industrial engineering and engineering management (IEEM)*, Macao, China, 15–18 December 2019 (pp. 1314–1318).

GREENSTEEL for Europe. In: Climate-neutral steelmaking in Europe decarbonisation pathways—Investment needs—Policy conditions—Recommendations. Written by Hien Vu, Francesca Cecchin. Edited by Nadina Iacob, Cristian Stroia, European Commission, November—2021, Brussel.

Herzog, K. W. (2017). The digitalisation of steel production. *BHM Berg - und Hüttenmännische Monatshefte, 162*(11), 504–513.

Herzog, K., Winter, G., Kurka, G., Ankermann, K., Binder, R., Ringhofer, M., Maierhofer, A., & Fick, A. (2017). The digitalisation of steel production. *BHM Berg- und Hüttenmännische Monatshefte, 162*, 504–513. https://doi.org/10.1007/s00501-017-0673-9

Iannino, V., Colla, V., Denker, J., & Göttsche, M. (2019). A CPS-based simulation platform for long production factories. *Metals, 9*, 1025. https://doi.org/10.3390/met9101025

Industry 5.0: Towards a sustainable, human-centric and resilient European industry (7 Jan. 2021). European Commission, Brussels. Retrieved from https://op.europa.eu/en/publication-detail/-/publication/468a892a-5097-11eb-b59f-01aa75ed71a1/

Kagermann, H., & Winter, J. (2018). The Second Wave of Digitization (How the Internet of Things, Industrie 4.0, Autonomous Systems, and Digital Business Models change the world), In S. Mair, D. Messner, & L. Deyer (Eds.), *Germany and the World 2030. What will change. How we must act* (pp. 201–209). Berlin: ECON Publishers.

Kiessling, S., Darabkhani, G., & Soliman, A. H. (2022). The bio steel cycle: 7 Steps to net-zero CO2 emissions steel production. *Energies, 15*, 8880. https://doi.org/10.3390/en15238880

Kusiak, A. (2017). Smart manufacturing. *International Journal of Production Research, 56*(1–2), 508–517. https://doi.org/10.1080/00207543.2017.1351644

Li, Y. R., & Li, T. K. (2021). Architecture based on big data for iron and steel intelligent manufacturing systems. *Metal Industrial Automation, 45*, 1–8.

Liu, W. Z. (2020). Current situation and thinking of intelligent manufacturing in China's iron and steel industry. *China Metallurgy, 30*, 1.

Mc Kinsey&Company. (2016). Supply chain 4.0—The next-generation digital supply chain (October 27, 2016). Retrieved from https://www.mckinsey.com/capabilities/operations/our-insights/supply-chain-40--the-next-generation-digital-supply-chain/

McKinsey&Company, Mori, L., Saleh, T., Sellschop, E., & Van Hoey, M. (2018). Digital opportunity map for a metal payer. In: Unlocking the digital opportunity in metals. Metals and Mining Practice (p. 4).

Miśkiewicz, R., & Wolniak, R. (2020). Practical application of the industry 4.0 concept in a steel company. *Sustainability, 12*, 5776. https://doi.org/10.3390/su12145776

Muslemani, H., Liang, X., Kaesehage, K., Ascui, F., & Wilson, J. (2021). Opportunities and challenges for decarbonizing steel production by creating markets for 'green steel' products. *Journal of Cleaner Production, 315*, 128127. https://doi.org/10.1016/j.jclepro.2021.128127

Neef, C., Hirzel, S., & Arens, M. (2018). Industry 4.0 in the European Iron and Steel Industry: Towards an Overview of Implementations and Perspectives. *Engineering, Materials Science, Business*. Corpus ID: 221082397, https://www.isi.fraunhofer.de/content/dam/isi/dokumente/cce/2018/Industry-4-0-Implementation-and-Perspectives_Steel-Industry_Working%20document.pdf.

Net Zero Tracker. (2022). Net Zero Stocktake 2022. Cologne, Berlin, Oxford, London, North Carolina, Germany, UK, United States of America: NewClimate Institute, Oxford Net Zero, Energy & Climate Intelligence Unit; Data-Driven EnviroLab. Retrieved July 14, 2022, from https:// ca1-nzt.edcdn.com/Net-Zero-Tracker/Net-Zero-Stocktake-Report-2022.pdf?v=1655074300

Peters, H. (2016). Application of Industry 4.0 concepts at steel production from an applied research perspective. Presentation at *17th IFAC symposium on control, optimization, and automation in mining, mineral and metal processing*. Retrieved from https://tc.ifac-control.org/6/2/files/symposia/vienna-2016/mmm2016_keynotes_peters.PowerPoint-Präsentation (ifac-control.org).

Peters, H. (2017). How could Industry 4.0 transform the steel industry? Presentation at *Future Steel Forum, Warsaw*, 14–15.6.2017. Retrieved from https://futuresteelforum.com/content-images/speakers/Prof.-Dr-Harald-Peters-Industry-4.0-transform-the-steel-industry.pdf

Pike, J. (2017). Innovative uses of hydrogen in steelmaking. <Midrex> 2017. Retrieved March 16, 2023, from https://www.energy.gov/sites/prod/files/2017/05/f34/fcto_may_2017_h2_scale_wkshp_ripke.pdf

Podsiadło, J., Dzienniak, S., & Zagórska, M. (2023). Megatrendy wpływające na branżę stalową. In: Stan i perspektywy hutnictwa żelaza i stali w Polsce. Nowe wyzwania. Monograph, (Ed.) Podsiadło J. Publisher: Sieć Badawcza Łukasiewicz – Górnośląski Instytut Technologiczny, Gliwice/Katowice 2023 (pp. 23–30).

Portal: ZŁOM INFO PL. Dekarbonizacja przemysłu stalowego - nadchodzi era „zielonej stali". 7 Dec. 2021. Label: dekarbonizacja hutnictwa#co2#zielona stal. https://www.zlom.info.pl/aktualnosci-informacje-z-branzy-zlomu/4434/dekarbonizacja-przemyslu-stalowego-nadchodzi-era-zielonej-stali.html

Rutkowski, K. (2004). Zarządzanie łańcuchem dostaw – próba sprecyzowania terminu i określenia związków z logistyką. *Gospodarka Materiałowa i Logistyka, 12*, 2–8.

Santos, C., Mehrsai, A., Barros, A. C., Araújo, M., & Ares, E. (2017). Towards Industry 4.0: An overview of European strategic roadmaps. In *Manufacturing engineering society international conference, MESIC 2017*, 28–30 June 2017, Vigo (Pontevedra), Spain.

Shin, K., & Park, H. (2019). Intelligent manufacturing systems engineering for designing intelligent product-quality monitoring system in the Industry 4.0. In *Proceedings of the 2019 19th international conference on control, automation and systems (ICCAS 2019)*, Hokkaido, Japan, 6–9 July 2019 (pp. 1693–1698).

Torn, I. A. R., & Vaneker, T. H. J. (2019). Mass Personalization with Industry 4.0 by SMEs: A concept for collaborative networks. *Procedia Manufacturing, 28*, 135–141.

Yin, R. Y. (2017). A discussion on intelligent steel plant-view from physical system side. *Iron Steel, 52*, 1–12.

Yao, L., & Wang, J. S. (2020). Goal and realization of intelligent manufacturing in steel industry. *China Metallurgy, 30*, 1–4.

Wang, B., Tao, F., Fang, X., Liu, C., Liu, Y., & Freiheit, T. (2021). Smart manufacturing and intelligent manufacturing: A comparative review. *Engineering, 7*, 738–757. https://doi.org/10.1016/j.eng.2020.07.017

Wu, L., Yue, X., Jin, A., & Yen, D. C. (2016). Smart supply chain management: A review and implications for future research. *The International Journal of Logistics Management, 27*(2), 395–417. https://doi.org/10.1108/IJLM-02-2014-0035

World Steel Association. Smart manufacturing. Retrieved November 20, 2023, from https://worldsteel.org/about-us/smart-manufacturing/

World Steel Association. (2020). Steel statistical yearbook 2020 concise version. Brussels: World Steel Association. Retrieved June 8, 2021, from https://www.worldsteel.org/en/dam/jcr:5001dac8-0083-46f3-aadd-35aa357acbcc/Steel%2520Statistical%2520Yearbook%25202020%2520%2528concise%2520version%2529.pdf

World Steel in Figures. (2023). World Steel Association. Retrieved November 20, 2023, from https://worldsteel.org/wp-content/uploads/World-Steel-in-Figures-2023-1.pdf

Zhang, G., Yang, Y., & Yang, G. (2023). Smart supply chain management in Industry 4.0: The review, research agenda and strategies in North America. *Annals of Operations Research, 322*, 1075–1117. https://doi.org/10.1007/s10479-022-04689-1

Zheng, Z., Zhang, K., & Gao, X. (2022). Human-cyber-physical system for production and operation decision optimization in intelligent steel plants. *Science China Technological Sciences, 65*, 247–260.

Zhou, D., Xu, K., Lv, Z., Yang, J., Li, M., He, F., & Xu, G. (2022). Intelligent manufacturing technology in the steel industry of China: A review. *Sensors, 22*, 8194. https://doi.org/10.3390/s22218194

CHAPTER 3

The Polish Steel Industry in the Context of Industrial Transformation

Abstract The Polish steel industry has only been on the market for three decades. Poland's political transformation began in 1989, at which time the government decided that the economy from a centrally controlled one would be a free market economy. The steel sector was in poor economic, technological and operational condition. In 2001, already during Poland's accession to the EU, the Polish government adopted steel industry restructuring programmes, which opened the Polish steel market to foreign investors and created the background of the steel market in Poland. Key changes, in the period from 1989 to 2022, in the Polish steel sector are shown in this chapter.

Keywords Polish steel industry · Steel market · Restructuring changes

3.1 Polish Steel Market

The economic transformation in Poland began in 1989, when the Polish government (under pressure from society) decided that the economy from centrally controlled would change into a market economy. The steel sector in Poland at that time (in the 1990s) was in a bad economic condition. The 1970s and the beginning of the 1980s were a good period for the Polish steel industry. In 1980, steel plants in Poland smelted 19.5

B. Gajdzik, *Digital Transformation towards Smart Steel Manufacturing*, https://doi.org/10.1007/978-3-031-77756-1_3

million tonnes of crude steel, so more than at that time was produced by the United Kingdom, India and Brazil. At that time, the Polish steel industry ranked 8th in the world. In 1989, steel production fell to 15 million tonnes, and the position in the world ranking was 13.

In 2001, during accession talks (Poland's accession to the EU), the Polish government accepted an important Steel Industry Restructuring Program. The Program included a provision to merge Poland's largest steel mills into a single organisation and provide it with state aid to conduct the transformation. Since the state aid, according to EU rules, had to be accepted by the European Commission (EC), discussions took place throughout 2002 as part of the accession negotiations to establish a mutually acceptable restructuring programme. In the restructuring process, the European Commission accepted the granting of state aid to the Polish steel industry up to a maximum of more than PLN 3 billion, but in return demanded a reduction in production capacity by 1 million tonnes of finished steel products (Podsiadło et al., 2023). The negotiated restructuring directions were included in Protocol No. 8 to the Accession Treaty. The steel sector in Poland was changing, at the time, according to the guidelines of the Treaty.

The restructuring programme of the Polish steel industry, under EC control, was finished in 2007. In the transformation time, large steelworks in Poland were sold to foreign investors. In 2002, through the consolidation of four steelworks: Katowice, Sendzimira, Cedler and Florian, the first consortium called "Polskie Huty Stali" (PHS) was formed. In 2003, the company was taken over by the British strategic investor LNM (Mittal). The buyer set up a company called ISPAT. After a few years, the company changed its name to Mittal Steel Company. Since 2006, the company has been part of the ArcelorMittal Group. In addition to ArcelorMittal, Spanish Celsa—owner of Huta Ostrowiec and American CMC Group—owner of Huta Zawiercie are in the Polish steel industry (Gajdzik & Sroka, 2012). Foreign capital groups in the Polish steel market are among the world's leading producers. The largest companies in the Polish steel market belong to ArcelorMittal (ArcelorMittal Poland), CMC (CMC Zawiercie), Celsa (Celsa Huta Ostrowiec, Alchemia). Among these groups in the Polish steel market, ArcelorMittal (AM) is the largest. In 2022, the Polish steel industry by volume of production took the twenty-third position in the world ranking according to the World Steel Association. On the market of steel producers in the European Union, Poland took sixth place (World Steel, 2023).

In the 1990s and later, the restructuring process of the Polish steel industry was mandatory (legal obligation) with a wide range of ownership, property, technological changes and employment reductions (Gajdzik, 2012a, 2012b)·. During that period, inefficient production facilities were dismantled, employment was drastically reduced, a number of production lines were modernised and new facilities were built, e. g. Hot Plate Rolling Mill in Cracow with a production potential of 5 million tonnes per year. In 2002, the open-hearth furnace process was stopped, and since then, two technologies have been used, these technologies are EAF and BOF. As a result of completed investment projects (primarily a consequence of privatisation processes), which have exceeded PLN 14.5 billion since Poland's accession to the European Union, Poland's steel industry has become one of the most modern in the world. Poland's membership in the European Union (as of May 01, 2004) has opened new opportunities for the steel industry in Poland, but it has also resulted in the associated necessity to implement European policies including climate policy, green energy, digital business, steel skills 4.0, etc. In the last two technological revolutions, people are increasingly being replaced by machines, robots, autonomous computer system etc. Of the total number of people employed in the Polish steel sector (1980s) of 149,000 persons, today 24,000 people work in the industry. Moreover, the structure of employees has changed: fewer blast furnace workers, rolling mill workers and more managers, operators and maintenance workers, engineers, IT specialists and others (Gajdzik, 2021).

After the restructuring, the steel mills are modern, with high productivity and job security, competitive and profitable. Changes based on reduction of steel capacity and employment combined with process automation yielded results. Steel mills allowed to achieve high productivity. The process of change has not stopped, now the steel mills are developing towards Industry 4.0 (Gajdzik & Wolniak, 2021a, 2021b). This process of steel transformation is strongly laced with climate policy and deep decarbonisation of steel production (Gajdzik et al., 2023).

By the end of the 1980s, the Polish smelters were able to produce more than 15 million tonnes of steel, in the following years production declined. In the 1990s (period from 1994 to 1997) production was over 10 million tonnes of steel, and even 11 million tonnes of steel per year. In the following years, high steel production in Poland (over 10 million tonnes of steel per year) was, in 2004, 2007, 2017 and 2018. During the pandemic year in 2020, steel production in Poland fell to

7.8 million tonnes of crude steel, a decrease of 13% compared to 2019. After the COVID-19 pandemic in 2021, 8.5 million tonnes of steel were produced in Poland. In 2022, Poland reduced steel production by 9% compared to 2021—to 7.4 million tonnes. Thus, the country was ranked 22nd in the global steel producer ranking by WorldSteel. Low production volumes were also recorded in Poland during the world economic crisis in 2009. Steel production in that year amounted to 7.1 million tonnes of crude steel. Details of annual steel production in Poland are presented in Fig. 3.1. The average annual production from 1989 to 2023 was 9.66 million tonnes. In 2023, Polish steel mills reduced steel production to 6.4 million tonnes.

In 2002, the last time in Poland, steelworks used OHF technology. Steel production in the technology in 2002 was 7,000 tonnes. For comparison, in 1989, 5.4 million tonnes of steel were produced in OHF process (36% of total production). OHF technology was uneconomical and incompatible with environmental standards. When steelworks were restructured, more uneconomical technologies were eliminated. Since 2003, the Polish steel industry has used two technologies: blast furnaces and converters (Blast Furnace + Basic Oxygen Furnace, abbreviation: BF + BOF) and Electric Arc Furnace (abbreviation: EAF). One smelter has the BF-BOF technology, and others have electric furnaces. Result of both processes is total steel production in Poland.

As part of the technological restructuring, COS (abbreviation in Polish language) was used in the steelworks. In the 1990s, the share of continuous steel casting (COS) in the total production was at the level of 10%. Now, total production is realised in technological sequences (Fig. 3.2).

There are unused production capacities in the Polish steel industry. Over the last two decades, the sector has been running on average about 70% of production capacity per year (excluding three last years). The reality of increasing steel production is determined by the high apparent consumption of steel in Poland. Since 2005, in the Polish steel market, the volume of domestic production has been less than the demand of the national economy for steel products. Figure 3.3 shows the value of domestic production against production capacity (a strong decline in 2020 due to the liquidation of the raw material section at Poland's largest steel mill) and the dynamics of consumption of steel products.

In the Polish steel consumer market, there are producers and companies from the following sectors: metal products, construction, automotive, machinery and equipment, household appliances, electrical appliances,

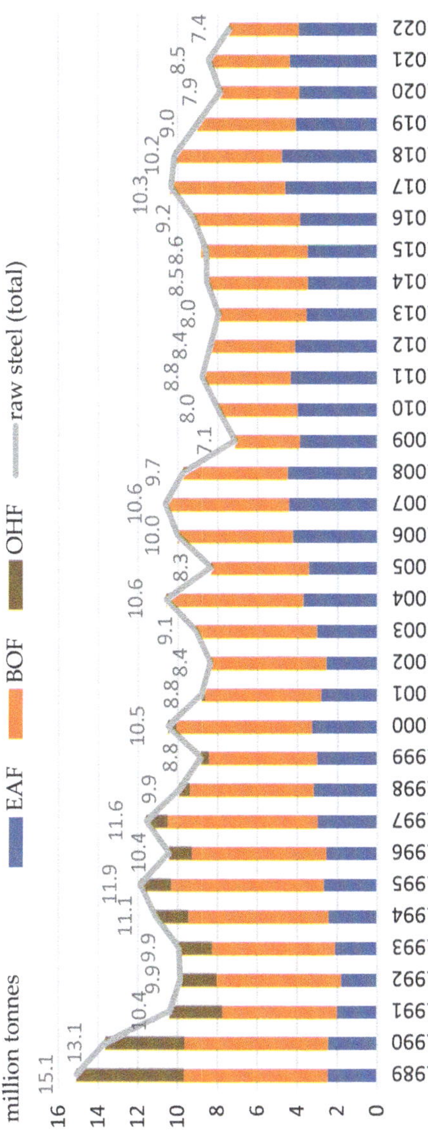

Fig. 3.1 Steel production in Poland from 1989 to 2022 (based on data of WorldSteel reports)

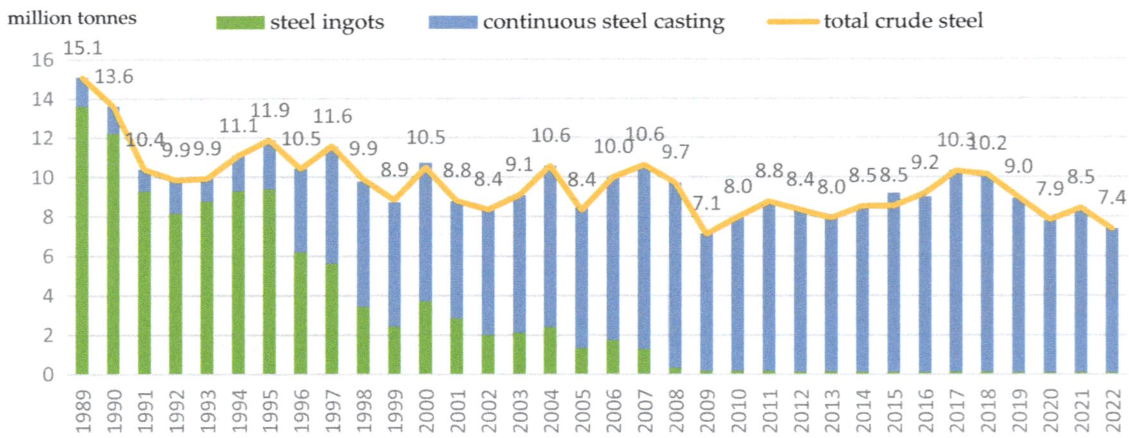

Fig. 3.2 Process of continuous steel production in Poland from 1989 to 2022. Source: Own elaboration based on data from Polish Steel Association in Katowice

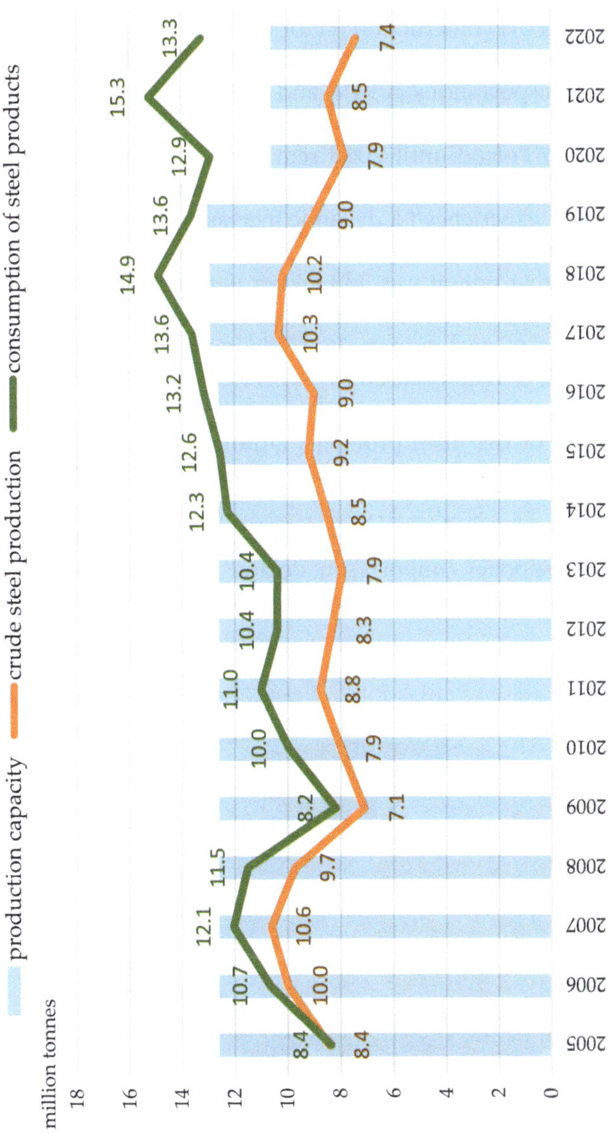

Fig. 3.3 Steel production and steel consumption in Poland from 2005 to 2022. Source: Own elaboration based on data from Polish Steel Association in Katowice

transport equipment, electronics (Fig. 3.4). In direct consumption, a high share belongs to the producers of metal products (39%). The key external customer is the construction sector (41%). The next places, in the steel consumer market belong to: automotive industry (14%), machinery industry (11%), transport industry (4%) and household appliance industry (3.4%) (Gajdzik & Zagórska, 2020).

Steel intensity in Poland until 2013 remained below the level of steel intensity for the EU, with an average rate of 265 kg per capita per year. After 2013, Poland experienced a dynamic increase in steel intensity, with the national steel intensity in 2018 reaching 390 kg per capita. By comparison, in the EU at that time it was 333 kg per person, and globally it was 225 kg per person. In the period 2020–2021, steel intensity in Poland decreased to 368 kg of steel per person, while in the EU it fell to 308 kg per person (Gajdzik & Zagórska, 2020). This trend continued in 2022 as well.

The assortment of the main products, although remaining the same, changed in production volumes especially flat products (a decline after 1990)—Table 3.1.

In foreign trade of steel products, since 2005, the volume of steel products from foreign supplies (imports) in Poland's steel consumption was higher than domestic supplies (domestic production) (Gajdzik, 2020).

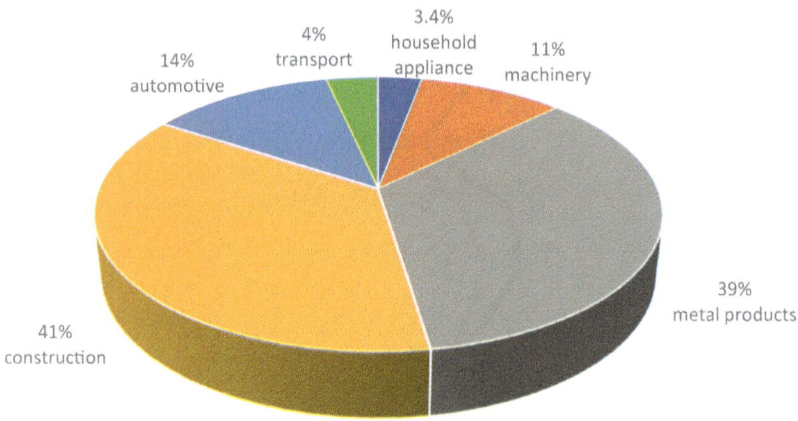

Fig. 3.4 Consumers of steel products in Poland. Source: Own elaboration based on data from Polish Steel Association in Katowice

Table 3.1 Main steel products in Poland (million tonnes)

Steel products	*1990*	*2000*	*2005*	*2010*	*2015*	*2022*
Total, including	10.453	7.470	6.188	6.894	7.831	7.215
Long products	4.400	4.422	3.917	4.545	4.943	4.929
Flat product	6.053	3.049	2.271	2.349	2.888	2.286
Cold-formed closed tubes and profiles	0.566 (without steel profiles)	0.805	0.742	0.950	0.911	0.647

Source: Own elaboration based on data from Polish Steel Association in Katowice

The steel industry records a negative, in terms of volume and value, balance of trade with foreign countries. This trend in foreign steel trade is shown in Fig. 3.5.

The next analysed area was employment, which radically (strongly) declined in the Polish steel sector after 1990. The highest level of employment in the Polish steel sector was in the 1989s, with over 150

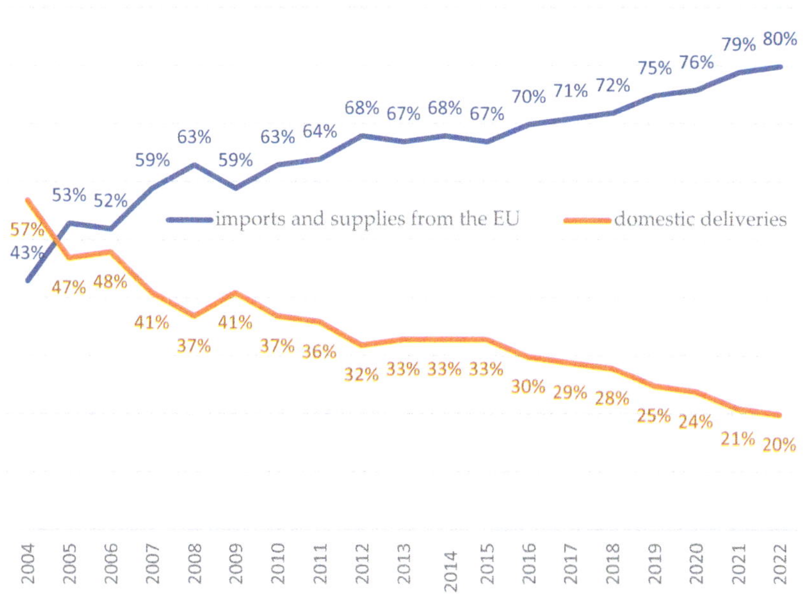

Fig. 3.5 Foreign trade of steel products in Poland from 2004 to 2022. Source: Own elaboration based on data from Polish Steel Association in Katowice

thousand people. In the nineteen nineties, 3.6% of all industry workers were employed in the metallurgy industry. 38,661 people were employed as nonproduction workers which accounted for 26.3% of all employees. Production efficiency was low, 125 tonnes of steel were produced per 1 worker. The obtained rate was 3 or more times lower than productivity levels in steel mills in Europe (Gajdzik & Ocieczek, 2015). During the transformation of steel mills to market economy conditions, there was a downward trend in employment in the Polish steel industry (Fig. 3.6). Over the past decade, employment in the steel industry has remained well below 25,000 (in 2021, 23,570; in 2022, 23,600), which means supporting as many as 154,000 jobs in the Polish economy (based on direct, indirect and induced effects). The share of directly employed in the steel sector in the average employment of the business sector in manufacturing is 1.0%, and in the economy is 0.4%. The total share of the domestic steel industry, combined with jobs in related sectors, accounts for about 2.4% of total employment in the national economy. After the restructuring process, production efficiency is comparable to that in the EU (over 300 tonnes of steel per 1 person).

The transformation of the Polish steel sector must also show the energy intensity of steel production. Energy consumption in the Polish steel industry is more than eight times higher than the average for all industries in Poland. The average annual consumption of electricity in the Polish steel industry per tonne of steel is 0.2 toe. There is more energy consumption in electric steel plants than in steel plants with blast furnace and converter technology. In the electric process the unit consumption of electricity is 3.5 times higher than in the integrated process (blast furnace and converter technology BF-BOF), where the main energy products are coke and process gases (Dzienniak, 2020)· In the category of total resources consumed in steel production, electricity is in second place with a share of 19%. The first position in the overall structure of energy resources consumed is occupied by coke and coal with a share of 47%. The third position belongs to electricity (18%). Electricity and coke together account for 65% of the total energy media used in the Polish steel industry (Zdonek et al., 2000, p. 31). The remaining positions are occupied by natural gas, blast furnace gas, coke oven gas and other utilities (Fig. 3.7). In the energy transformation that is taking place in the Europe, Poland's economy must increase the share of renewable energy in total energy production, and the various energy-intensive industries, including steel mills, must diversify their energy sources. In this transformation, steel

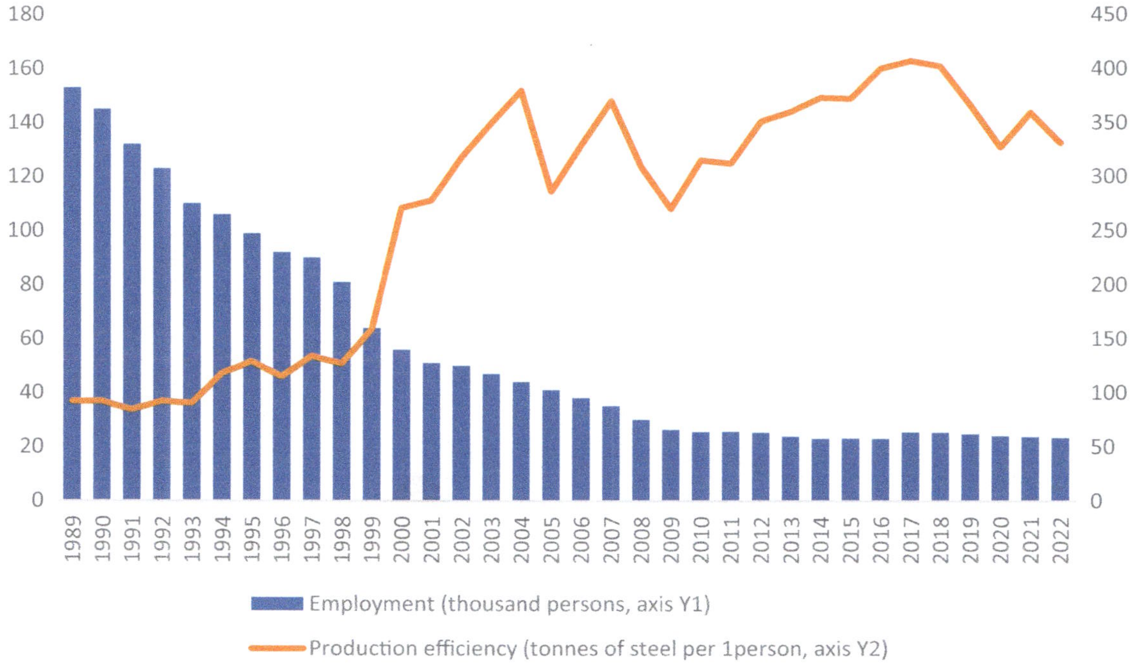

Fig. 3.6 Employment and production efficiency in the Polish steel market from 1989 to 2022. Source: Own elaboration based on data from Polish Steel Association in Katowice

mills will strive, on the one hand, to introduce green hydrogen and other green energy sources, and on the other hand, to invest in technologies that consume less energy than existing ones (Gajdzik & Sroka, 2021; Gajdzik et al., 2021; Wolniak et al., 2020).

The annual average energy consumption per tonne of steel (0.293 toe) in the Polish steel industry is a little lower than the average consumption in the European Union, where energy intensity is 0.312 toe/tonne of steel. Figure 3.8 presents the energy intensity of production in Poland based on data from Polish Statistics (GUS).

Steel mills are investing in modern technologies that increase the share of process (blast furnace and coke oven) gases (Borowiecki & Wysłocka, 2003). From the 1990s to now, this sector has spent about PLN 23 billion on technological investments (Fig. 3.9). For many years now, investments in steel mills have been strongly linked to higher energy efficiency in steelmaking processes. Authors Wolniak et al. (2020) confirmed in a study that an increase in investment spending by 1 million zlotys results in a decrease in energy consumption by 0.0000357 toe/t. Similar conclusions were reached by Gajdzik and Sroka (2021), analysing

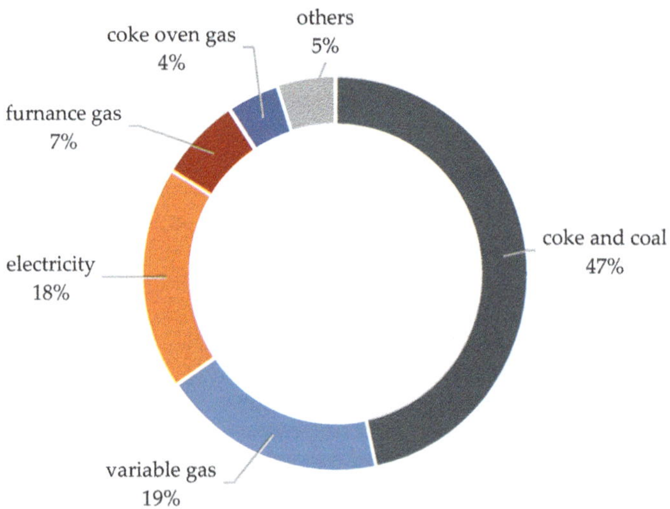

Fig. 3.7 Energy resources in the Polish steel market. Own elaboration based on data from Polish Steel Association in Katowice

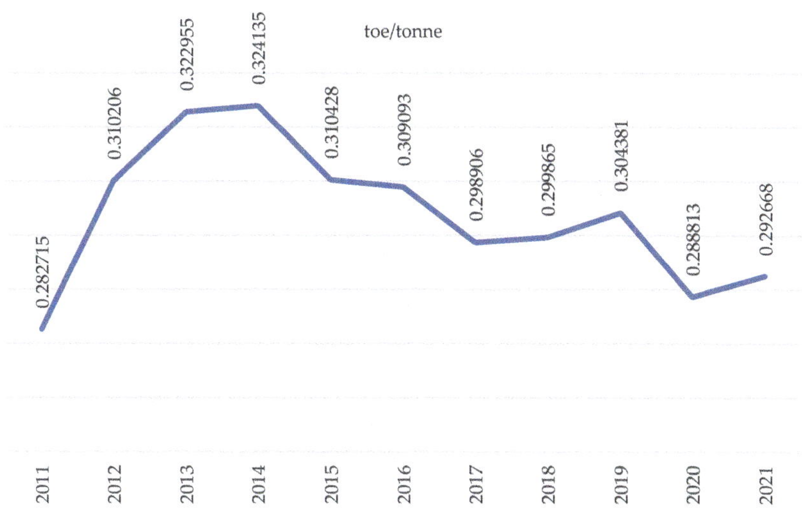

Fig. 3.8 Energy consumption per 1 tonne of steel in Poland from 2011 to 2021. Source: Own elaboration based on data from Statistics Poland (GUS) Warsaw, reports: Energy efficiency in Poland

resource intensity in terms of coke and energy consumption in steel mills. A decrease in resource intensity contributes to a decrease in unit production costs in the steel industry.

Among the investments made were steel smelting technologies (in 2003, Siemens–Martin process was stopped—technology discontinued) (Gajdzik, 2011, 2012a, 2012b) and continuous steel casting technologies. These investments were accompanied by ecological innovations, which were caused by the need to adapt the Polish steel industry to the requirements of the European market (legislation on limiting emissions of pollutants in European Union countries). In one year alone, 2007, steel mills spent PLN 2 billion on investment (steel industry realised the European programme based on viability). In last years, steel mills have invested in automation and ICTs. Steel production processes are continuously digitised. Computer systems monitor the manufacturing operations and control the technological and processes parameters (computer systems control the processes: steel melting, rolling and others, and intelligent sensors provide real-time data). Key technologies, including steel smelting, are fully automated. Equipment in steel mills

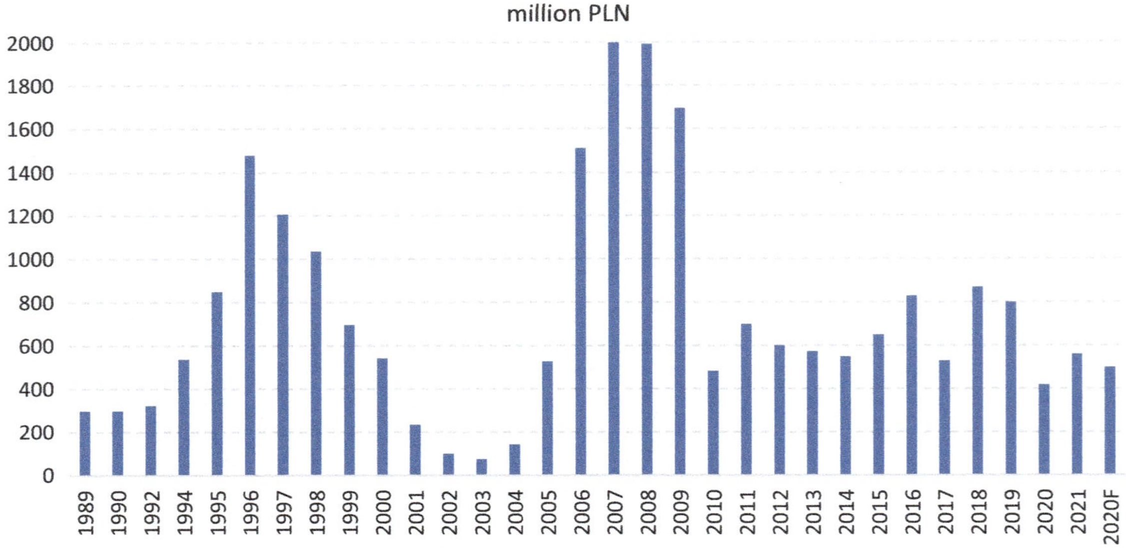

Fig. 3.9 Expenditures on investments in steel industry in Poland from 1989 to 2022. Own elaboration based on data from Polish Steel Association in Katowice

and rolling mills is being equipped with intelligent sensors that provide real-time process data. Huge advances have been made in maintenance and machine management (TPM) on the level of autonomous (AM) and professional (PM) (more in: Gajdzik, 2014). The Polish steel industry manage processes according to pillars of World Class Manufacturing (WCM) (Gajdzik, 2013a, 2013b). Moreover, with the popularisation of the Industry 4.0 concept, steel plants invest in additive manufacturing, including 3D printing, AI with machine learning, HMI (human and machine interface), augmented reality (AR), etc. (Gajdzik, 2022). The key realised investments in the Polish steel industry are presented in Table 3.2.

3.2 Polish Steel Industry in European Steel Market

Until 1989, when there was a government-controlled central economy in Poland and no market economy, the Polish steel sector was strongly linked to the supply chain of the socialist side countries, which included many Central and Eastern European countries. Trade in steel and steel products took place within a group of countries linked by economic systems. As late as the 1980s, the Union of Soviet Socialist Republics (USSR, in Russian: CCCP) and countries with industrial ties to this powerful country produced more than 30% of the world's steel output (Gajdzik, 2013a, 2013b). During the 1990s (the period from 1990 to 1993), steel production in the USSR, together with countries politically and economically linked to the USSR, fell from 203 million tonnes to 128 million tonnes of steel. This was a larger decline in the history of the development of this bloc of countries. Along with the decline in production, the consumption of steel products in the bloc of socialist countries declined steadily. In 1990, the consumption of steel products in these countries amounted to 145 million tonnes of steel. Three years later (1993), steel consumption in these countries was 82 million tonnes of steel (Szulc & Kardas, 2009).

In the 1990s, many countries (Poland, Hungary, the Czech Republic, Romania, Slovakia) of the former socialist bloc, proceeded to restructure their industries, including steel mills. In addition, the new countries, including Poland, signed the Europe Agreement (on accession), so steel mill repair programmes were agreed with the European Commission. The steel industry in Poland and other Central and Eastern European countries began the process of transforming steel mills into market economy

Table 3.2 Key realised investments in the Polish steel industry

Plant/steelworks	*Investment/place/year*	*Expenditure*
ArcelorMittal Capital Group	Hot plate rolling mill in Cracow /June 2007	PLN 1.159 billion
	Continuous steel casting line (No. 3)/ capacity: 3 million tons of slabs per year Dąbrowa Górnicza /December 2006/	PLN 447 million
	Modernisation of the rolling plant/ Sosnowiec/November 2006	PLN 140 million
	Sheet metal coating line/ plant in Świętochłowice / September 2006	PLN 107 million
	Modernisation of blast furnace (No. 2)/ Dąbrowa Górnicza / production capacity 2.7 million tons per year / October 2006	PLN 260 million
	New coke oven batteries (two)/ Zdzieszowice /2007—Battery No. 11, 2008—Battery No. 12	PLN 364 million
	Two-station vat furnace with technology for automotive grades/capacity: 4.5 million tons per year/ Dąbrowa Górnicza /June 2009	PLN 132 million
	Hot rolling/Cracow/after 2010	PLN 230 million
	Blast furnace/Cracow/ after 2010	PLN 175 million
	Galvanising line / Cracow/ after 2010	PLN 95 million
	Coal-mining department/ Zdzieszowice plant/after 2010	PLN 205 million
	Line for long rail products/ Dąbrowa Górnicza/ after 2010	PLN 140 million
	Other investments/ecological/Dąbrowa Górnicza	PLN 700 million
Celsa/Ostrowiec	Rolling mill for long products/ new products for the energy sector Steel plant/furnace	PLN 3 billion
CMC Poland/ Zawiercie	Technology for shredding and cleaning of steel scrap Modern rolling lines/2009/2010 Electric furnace (EAF)/2014 Continuous steel casting line/2016 Rolling mill/2021	PLN 2 billion
ISD Steelworks Częstochowa	Modernization of steel production lines	PLN 1.7 billion

Source: own elaboration based on Podsiadło et al. (2023)

(Gajdzik, 2013a, 2013b). During the period of steel industry restructuring, the 1990s, steel mills in the century of countries that changed the economic system were given state aid and their debts were written off by their governments. Some countries applied to the EC for an extension of the period of financial support for restructured steel mills. According to the European Commission's viability guidelines, the steel mills in Central and Eastern European countries achieved the financial indicators set for them. Unfortunately, at that time—the restructuring of the steel industry in the CEE countries—many steel mills were liquidated, and those that survived the difficult period were incorporated into foreign capital groups, such as Nova Huta Ostrava (Czech Republic)—Mittal, Polskie Huty Stali (Poland)—Arcelor Mittal, Galati (Romania)—ArcelorMittal, Huta Koszyce (Slovakia_ US Steel, Hunedoara (Romania—ArcelorMittal (Gajdzik, 2013a, 2013b).

In the global steel market, in the 90 s and the following years, there was a consolidation of capital, then the largest capital groups emerged—global steel producers. The consolidation of the global metallurgy sector started from the merger of Krupp and Hoesche in the 90 s of the previous century. In the 90 s in Europe, there were over 120 transactions of consolidation on a smaller scale and only in 1997, there were 5 of them. As a result of consolidation such concerns as Corus (merger of British Steel—UK and Hoogovens—Holland), ThyssenKrupp (merger of Thyssen and Krupp), Arcelor (French Usinor, Luxembourg's Arbed and Spanish Aceralia) were created. Consolidation processes occurred also in Northern America and on the wave of the bankruptcies of many small steel producers a group called International Steel Group (ISG) was formed and in 2006 was taken over by Mittal Steel. In Japan, due to the merger of Kawasaki Steel and NKK a concern called JFE was created in 2003. In 2006, the British Corus was taken over by Indian concern Tata Steel. An important consolidation for the Polish steel market was the creation of the ArcelorMittal group, which has a major position in the Polish steel industry. ArcelorMittal has its assets in Poland. The company concentrates about 70% of the country's production potential. The company includes branches in Dąbrowa Górnicza, Chorzów, Sosnowiec, Świętochłowice, Zdzieszowice and Cracow (Gajdzik & Sroka, 2012; Sroka, 2009).

The Polish steel market is a 97% private market and is based on steel production and processing. According to the Statistics Office in Poland more than 2000 companies (enterprises) create the section of economic activity—metals (Report, Statistical Yearbook of Industry Poland, 2022).

In the era of the market economy, steel mills in the age of Central and Eastern European countries adapted to the changes and were even able to compete with mills from countries that had long since reached market maturity. Table 3.3 presents the steel production in Central and Eastern European countries. In global ranking, the Polish steel industry (steel production in Poland) is in 23-rd position with 6.4 million tonnes steel in 2023.

Changes in the business and legal areas have meant that less steel has been produced in Poland in recent years, but metallurgy has a future. The Polish metallurgy (steel industry) is part of the European steel industry. The EU economy is an open economy, so, in recent years, is being increasingly displaced by imports from third countries. Until 2021, Poland consumed more steel (Table 3.4). Metallurgy is an important link in recycling for other industries, producing approx. 6.5 million tonnes of steel scrap. It is a decarbonised raw material, to which Poland's recovery makes a significant contribution. On the other hand, the European Union exports approx. 20 million tonnes of scrap, of which approx. from 1.5 to 2 million tonnes of this raw material is exported from Poland.

Steel companies in the Polish market produce a full range of steel products, and the distribution network in Poland is adapted to customer needs but there is a problem with demand fluctuations in the market. The Polish market has been affected by global events, including the COVID-19 pandemic and the war in Ukraine. During the pandemic, steel production

Table 3.3 Steel production in Central and Eastern European countries

County/year	*Steel production in million tonnes*								
	1995	*2000*	*2005*	*2010*	*2015*	*2020*	*2021*	*2022*	*2023*
Poland	11.890	10.498	8.336	7.993	9.198	7.856	8.454	7.407	6.437
Slovakia	3.975	3.733	4.485	4.583	4.562	3.444	4.863	3.872	4.470
Czech Republic	7.184	6.216	6.189	5.180	5.262	4.465	4.817	4.289	3.384
Hungary	1.861	1.871	1.958	1.678	1.675	1.513	1.100	0.857	0.476
Romania	6.555	4.672	6.280	3.721	3.352	2.790	3.375	2.625	1.580
Bulgaria	2.724	2.022	1.949	0.737	0.543	0.484	0.548	0.482	0.489
Croatia	0.045	0.071	0.073	0.095	0.122	0.045	0.185	0.169	0.220

Source: World Steel Association, Steel Statistical Yearbook

Table 3.4 Steel products total: Poland vs. the EU

Metric tonnes	*Y 2000*	*Y 2005*	*Y 2010*	*Y 2015*	*Y 2020*	*Y 2021*	*Y 2022*	*2023e*
Eu 27	15,206,088	30,682,461	29,439,016	38,114,250	33,929,998	44,424,858	40,547,982	36,448,437
Poland	271,174	1,055,385	1,209,750	2,242,538	2,769,190	3,966,667	2,702,330	2,346,484

Source: By CN8 Codes Imports Comext, steel products (total) CN 7206–7306, Third Out EU27

in Poland fell by 11.6% compared to the previous year, but the decline was not as strong as during the crisis that hit Poland in 2009, after the difficult situation on the real estate and banking market in the US, fell by 26.7% compared to steel production in Poland in 2008) (Gajdzik & Wolniak, 2021a, 2021b).

Many changes implemented in the metallurgical market result from regulations that have been introduced to achieve climate neutrality. Although per ton of steel produced in Poland, CO_2 emissions have fallen in recent years, the sector is still highly carbon intensive (Fig. 3.10) and energy intensity. Despite a profound technological restructuring in the first decade of the twenty-first century, the metallurgical industry is still one of the three most energy-intensive processing industries, with only 1TJ per year of final energy consumption coming from renewable sources (solid biofuels). Expensive investments in decarbonisation and an attempt to mass-scale certified production of "green steel" are necessary to return to the broad market. Scenarios of decarbonisation of the Polish steel sector concern two technologies: EAF and BF-BOF (Gajdzik et al., 2023). The first segment of producers (50% of Poland's annual total steel production) with EAF technology has problems related to high electricity costs caused by the high share of coal in Poland's energy production (the basic source of electricity is hard coal in Poland), as well as securing scrap supplies. The second segment—steel mills with BF-BOF technology, needs to invest in innovative technologies: DRI and CCU(U). With decarbonisation technology, however, there is a problem with obtaining clean hydrogen, and it does not seem to be solved soon. However, natural gas can be used instead of hydrogen, resulting in much lower CO_2 emissions than with the blast furnace process. In the proposed scenario, there may be a shift to producing steel from scrap in electric arc furnaces—EAF, but then more energy needs to be supplied to steel mills (Gajdzik et al., 2023). The purchase of electricity is between 8 and 30% of production costs for steel mills in Poland (such a large disparity comes from the fact that the share of electricity in costs is much higher in electric steel mills than in others). The energy crisis that emerged after the COVID-19 pandemic—high electricity prices—has affected many markets, including the steelmaking market. In October 2020, Poland's steelmaking branch in Cracow (AMP Steelworks) was shut down and Poland irreversibly lost 2 million tons of production capacity.

After 2020, the steel market in Poland, as well as in many EU countries, is very unstable. It is difficult to have predictability on it to manage

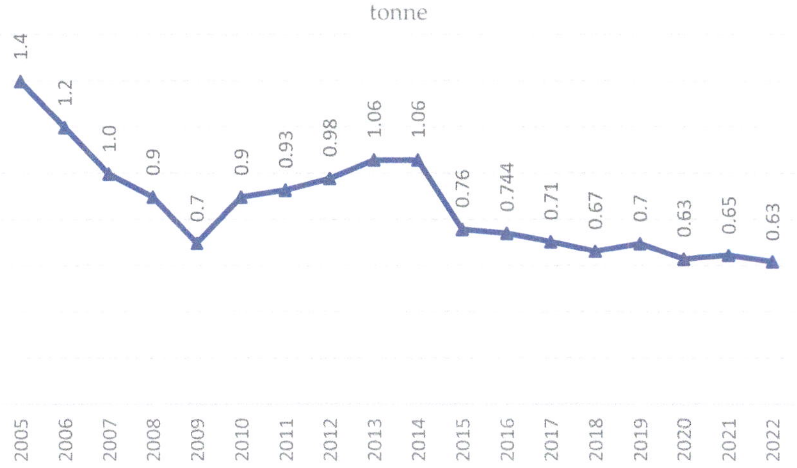

Fig. 3.10 CO_2 emissions per tonne of steel in Poland from 2005 to 2022. Source: own elaboration based on KOBiZE report obtained from Polish Steel Association, Poland, Katowice

economic processes effectively. First, it was hit by a pandemic, ripping up supply chains and then by the war in Ukraine. The Polish steel industry, at the threshold of the third decade of the twenty-first century, faces the need for technological modernisation in the not easy conditions of climate regulation and competition from non-EU steel producers. Opportunities for market development can be seen in new investments according to decarbonisation policies and trends in Industry 4.0.

In Industry 4.0, companies (steel producers) can create cyber-physical systems (CPS) with new generation of digital technologies. The typical digitalisation that preceded the Third Industrial Revolution no longer exists. In the Fourth Industrial Revolution, the new digitalisation is currently more autonomous. Companies links digital technologies with and physical objects and processes and create cyber-physical production systems (CPPS). In modern industrial transformation, the cyber-technologies and physical processes are becoming increasingly more intertwined. The innovative solutions can be opportunities for steel industry in global climate policy.

References

Borowiecki, R., & Wysłocka, E. (2003). Możliwości i kierunki optymalizacji kosztów produkcji w polskim hutnictwie. *Zeszyty Naukowe Uniwersytetu Ekonomicznego w Krakowie, Zeszyt, 618*, 16–33.

Dzienniak, S. (2020). Zużycie mediów energetycznych w hutnictwie. Hutnicza Izba Przemysłowo-Handlowa (HIPH), Katowice. In: *PTE conference*, Poland, Katowice, 22.09.2020, Ekonomiczne skutki pandemii. Prezentacja HIPH: Droga energia przyczyną niskiej konkurencyjności polskiej gospodarki na przykładzie przemysłu stalowego.

Gajdzik, B. (2013a). *Restrukturyzacja przedsiębiorstw hutniczych w zestawieniach statystycznych i badaniach empirycznych*. Silesian University of Technology.

Gajdzik, B. (2014). Autonomous and professional maintenance in metallurgical enterprise as activities within total productive maintenance. *Metalurgija, 53*(2), 269–272.

Gajdzik, B. (2021). Changes in employment and working conditions in the perspective of Industry 4.0. *Human Resources Management (ZZL), 1–2*, 11–26. https://doi.org/10.5604/01.3001.0014.8783

Gajdzik, B. (2022). Key directions in changes from steelworks 3.0 to steelworks 4.0 with analysis of selected technologies of digitalizing the steel Industry in Poland. *Management Systems in Production Engineering, 1*(30), 46–53. https://doi.org/10.2478/mspe-2022-0006

Gajdzik, B. (2012a). *Metallurgical plant after restructuring. Dynamics of changes in the domestic steel sector in the years 1992–2010*. Gliwice: Silesian University of Technology.

Gajdzik, B. (2011). The comprehensive analysis of restructuring changes in the iron and steel sector in Poland. In J. Pyka (Ed.), *Modernity of industry and services. Dynamics of changes in Polish industry and services*. TNOiK.

Gajdzik, B. (2012b). The road of Polish steelworks towards market success–changes after restructuring process. *Metalurgija, 51*(2), 265–268.

Gajdzik, B. (2013b). World class manufacturing in metallurgical enterprise. *Metalurgija, 52*(1), 131–134.

Gajdzik, B., & Ocieczek, W. (2015). Soft restructuring process in metallurgical enterprises in Poland. *Metalurgija, 54*(3), 729–732.

Gajdzik, B., & Sroka, W. (2012). Analytic study of the capital restructuring process in metallurgical enterprises around the World and in Poland. *Metalurgija, 51*(2), 265–268.

Gajdzik, B., & Sroka, W. (2021). Resource intensity vs. investment in production installations—The case of the steel industry in Poland. *Energies, 14*(2), 443. https://doi.org/10.3390/en14020443

Gajdzik, B., Sroka, W., & Vveinhardt, J. (2021). Energy intensity of steel manufactured utilising EAF technology as a function of investments made: The case of the steel industry in Poland. *Energies, 14*, 5152. https://doi.org/10.3390/en14165152

Gajdzik, B., & Wolniak, R. (2021a). Influence of the COVID-19 crisis on steel production in Poland compared to the financial crisis of 2009 and to boom periods in the market. *Resources, 10*, 4. https://doi.org/10.3390/resources10010004

Gajdzik, B., & Wolniak, R. (2021b). Transitioning of steel producers to the steelworks 4.0—Literature review with case studies. *Energies, 14*(14), 4109. https://doi.org/10.3390/en14144109

Gajdzik, B., Wolniak, R., & Grebski, W. (2023). Process of transformation to net zero steelmaking: Decarbonisation scenarios based on the analysis of the Polish steel industry. *Energies, 16*, 3384. https://doi.org/10.3390/en16083384

Gajdzik, B., & Zagórska, M. (2020). Analiza stalochłonność polskiej gospodarki przy użyciu wybranych mierników. Gosp. Mater. Logist., 72 nr 6.

Podsiadło J., Dzienniak S., & Zagórska M. (2023). Megatrendy wpływające na branżę stalową. In: Podsiadło J (ed.) Stan i perspektywy hutnictwa żelaza i stali w Polsce. Nowe wyzwania. Monograph, Publisher: Sieć Badawcza Łukasiewicz – Górnośląski Instytut Technologiczny, Gliwice-Katowice (pp. 23–30).

Report of Statistics Poland. (2022). https://stat.gov.pl/en/topics/statistical-yearbooks/statistical-yearbooks/statistical-yearbook-of-industry-poland-2022,5,16.html Table. 2., p.35, Economic entities by ownership sectors, sections and divisions (cont.).

Sroka, W. (2009). Competitive advantage through cooperation in networks on sector Level: The case of steel industry. In J. Kaluza (Ed.), *Strategic management and its support by information systems*. VSB – Technical University of Ostrava.

Szulc, W., & Kardas, M. (2009). Restrukturyzacja sektora stalowego w Europie Środkowo-Wschodniej. *Hutnik-Wiadomości Hutnicze, 76*, 149–154.

Wolniak, R., Saniuk, S., Grabowska, S., & Gajdzik, B. (2020). Identification of energy efficiency trends in the context of the development of Industry 4.0 using the Polish steel sector as an example. *Energies MDPI, 13*, 2867. https://doi.org/10.3390/en13112867

World Steel in Figures. (2023). World steel association, Brussels Belgium. https://unesid.org/descargas_files/World-Steel-in-Figures-2023.pdf

Zdonek, B., Mazur, A., Szypuła, I., & Stefaniak, A. (2000). Kierunki optymalizacji kosztów wsadowych, eksploatacyjnych i energii w procesie wytwarzania stali. Prace Instytutu Metalurgii Żelaza, 2, Gliwice (p. 31).

CHAPTER 4

Transitioning Towards Smart Steel Manufacturing: Results from Research in Poland

Abstract This chapter is about the changes towards Industry 4.0 in the steel market in Poland. In this chapter, the results of realised direct research on the Polish steel market are presented. The author of the study was PhD Eng. Bożena Gajdzik, assoc. Professor at the Silesian University of Technology. The study was preceded by a pilot study, the results of which were presented in the author's monograph, published by Silesian University of Technology in Gliwice (Diagnoza kierunków transformacji przemysłu stalowego w Przemyśle 4.0 (Diagnosis of the steel Industry transformation direction in the Industry 4.0). Silesian University of Technology, Gliwice, 2022). The research was carried out from 2020 to 2023 in steel market companies (enterprises) in Poland, including steel mills. The purpose of the research was to identify areas of transformation in the steel industry, taking into account positive and negative impact factors. For this purpose of the research, it was assumed that the transformation of Industry 4.0 is the carrier of the next wave of transformation through the displacement of traditional technologies (based on manual activities) by smart (high) technologies, which will form smart steel manufacturing. The impact of Industry 4.0, through its technological pillars, on changes in the steel industry was presented in this chapter.

Keywords Steel enterprises · Direct research · Smart transformation · Pillars of Industry 4.0

B. Gajdzik, *Digital Transformation towards Smart Steel Manufacturing*,
https://doi.org/10.1007/978-3-031-77756-1_4

4.1 Methodology of Research

The idea of research was born under the strong popularisation of the concept of Industry 4.0. This new concept, initiated in Germany in 2011, quickly gained publicity and became a strategic direction for the development of many industries. Industries have started their own paths to smart manufacturing. This transformation has also affected the steel sector. Peters (2017), Castelo-Branco et al. (2019), Gajdzik and Wolniak (2021) and others (Gajdzik, 2022; Gajdzik et al., 2025a, 2025b) assumed that the steel sector is undergoing a transformation to smart manufacturing. This assumption became the basis for my own direct research. This assumption became the basis for direct research carried out in the steel market in Poland. The research was realised by CAWI method. The structure of the research questionnaire was based on the nine pillars of Industry 4.0, described in many scientific publications (e.g. Senn, 2019; Lee, 2008; Erboz, 2017; Rüßmann et al., 2015; Büchi et al., 2020; Rojko, 2017, Dubey et al., 2022). To assess the industrial transformation of steel enterprises in Poland towards smart steel manufacturing, the Likert scale from 1 to 5 (where 1 was the lowest and 5 the highest) was used. The survey comprised 10 questions, which were subsequently categorised into four key areas of research. The questions of the survey focused on the following research areas (RA):

(RA_1) Respondents' knowledge of the concept of Industry 4.0 and its technological pillars.
(RA_2) The impact of Industry 4.0 on the smart transformation of the Polish steel industry.
(RA_3) The application of technological pillars of Industry 4.0 in the enterprises in the Polish steel industry.
(RA_4) Determinants of the smart transformation in the Polish steel market in the form of opportunities and barriers of this transformation.
(RA_5) Future directions of industrial transformation of the steel enterprises in Poland and expected progress of the smart transformation.

These research areas (RAs) were formulated based on the research questions. The matrix of research structure is presented in Table 4.1. The particular questions are presented in the subsection 4.2 of this chapter.

Table 4.1 Matrix of research structure: survey

Research Areas (RAs)	*Survey Questions (Qs)*
RA_1	Q_1, Q_2
RA_2	Q_3, Q_4
RA_3	Q_5, Q_6
RA_4	Q_7, Q_8
RA_5	Q_9, Q_10

Source: own elaboration

The research areas are analysed vertically (blue colour in Fig. 4.1) and horizontally (red colour in Fig. 4.1).

The analysis was performed using "R" computer program. Depending on the scope of the study, selected descriptive statistics and statistical tests were applied. The obtained statistical significance in the individual tests was the basis for formulating general conclusions of the research. The results of the study were presented in tables and figures in the form of boxplots and dendrograms.

According to Peters (2017), Castelo-Branco et al. (2019) and Gajdzik and Wolniak (2021), the scope of research is changes in steel mills and linked cooperative companies in steel supply chain. The purpose of research was to determine the technological scope of steel transformation in Poland. For this purpose of research, it was assumed that the transformation is the carrier of the next wave of transformation through the displacement of traditional technologies (based on manual activities) by smart (high) technologies towards smart steel manufacturing and flexibility of enterprises in the steel market.

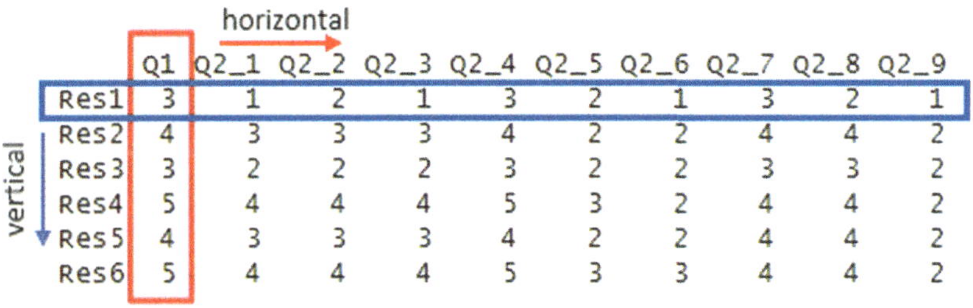

Fig. 4.1 The structure of research analysis. Source: own research

Based on this established research purpose, the following research questions (RQs) were formulated:

- RQ1: What technologies describe the industrial transformation of Polish steel enterprises towards smart steel manufacturing?
- RQ2: What are the determinants of smart transformation of enterprises in the steel industry in Poland?
- RQ3: What clusters of technological changes form the framework of the smart transformation in the steel industry in Poland?

The research was carried out according to the following stages:

Stage I: Preparation of the survey: Likert scale from 1 to 5, where 1 is the lowest, and 5 the highest was used in the survey.
Stage II: Validation of the questionnaire by experts of metallurgy.
Stage III: Selection of respondents according to formula (4.1).
Stage IV: Direct research (CAWI method).
Stage V: Analysis of research results (from aggregate analysis to cluster analysis).
Stage VI: Research conclusions.

The research was realised in the segment of steel enterprises in Poland. Each enterprise had one representative in this research. The survey involved 253 enterprises (managers or experts of changes in the enterprises). The pre-qualification of responders for the study was carried out in accordance with the requirements of the postulates of randomness and representativeness of the sample. The selection of the sample was confirmed by statistical tools (test in the computer program) according to the sampling methodology (Machin et al., 2009). The test was applied to determine the minimum sample size for estimating the structure index p in the general population: acceptable error d = 5%, initial assumption: the study population (a sample) of respondents is about 20% of the general population, initial estimate $p = m/n = 20\%$. The obtained test result of 245 was compared with the number of respondents of 253 survey participants for a confidence level of 95%, the obtained test result: $253 > 245$, and therefore the sample was sufficient for further research (the requirements of the postulates of randomness and representativeness of the sample were maintained). Selection of respondents for the study was realised based on their knowledge about the concept of Industry 4.0.

The following research questions were used:

Q_1 (Question 1): Do you have knowledge of the concept of Industry 4.0 (I 4.0)?
No, I have never heard of the concept = 0,
Yes, I have heard of the concept, but I don't know its characteristics = 2 points,
Yes, I know the pillars of the concept = 5 points,
Yes, I know more details about the concept = 8 points,
Yes, we are already implementing this concept in our enterprise = 10 points.
Q _2 (MEX: metrics): Expert knowledge of metallurgy and steel market:
Practical knowledge (many years of working for the enterprise) = 10 points,
Steel industry expert knowledge = 8 points,
Enterprise functions in the enterprises = 5 points,
others (experience, education, trainings, interpersonal relations) = 6 points.
Q_MEX_3 (MEX: metrics): Sources of knowledge of the concept of Industry 4.0
Cooperation with IT companies in the process of implementation of new solutions = 10 points,
Participation in scientific conferences and business meetings = 8 points,
Cooperation with universities and scientific institutes = 8 points,
Others: experience, education, trainings, interpersonal relations = 6 points.

The primary criterion for not qualifying a respondent as an expert was the answer to Q1: Do you have knowledge of the concept of Industry 4.0? Answer (Q1): No, I have never heard of the concept, in this case the expert competence coefficient (index) K_k took a value equal to zero and the respondent was excluded from the group of experts. If the point value of the answer to Q1 was different from zero, the expert competence coefficient (index) K_k in the modified expert method was determined based on the formula (4.1). On the basis of this analysis, 14 respondents were rejected. Finally, the statistical assessment of the survey (questionnaire) included 253 experts (respondents of steel enterprises).

$$K_k = \frac{\text{Analysis\_ AEX(points)} + \text{Analysis\_ MEX\_ 2(points)} + \text{Analysis\_ MEX\_ 3(points)}}{3} \quad (4.1)$$

Analysis_AEX(points for Q1); Analysis_MEX_2(points for Q_MEX_2) and Analysis_MEX_3(points for Q_MEX_3.

4.2 Results of Direct Research

RESEARCH AREA: RA_1

The first research area (RA_1) focuses on the respondents' knowledge of the concept of Industry 4.0. This is a significant issue as it indicates the degree of understanding and awareness among enterprises in the Polish steel sector regarding this modern industrial concept. This area of research was formed from two questions: Q_1. Do you know the concept of Industry 4.0? and Q_2. Do you have knowledge of the Industry 4.0 pillars (technologies)? Table 4.2 presents obtained answers. According to Q_1, respondents had one choice from the five-point Likert scale: I've heard of the concept and its pillars (1 point), I've heard of the concept and its pillars but I don't know the details (2 points), I've heard of the concept and can say something about it (3 points), I have a knowledge of the I 4.0 concept (4 points), I have a practical knowledge about the I 4.0 concept (5 points). The structure of the second question (Q_2) consisted of nine answers from Q2_1 to Q2_9 (Table 4.2). The answers (Table 4.2) corresponded to the pillars of Industry 4.0 described by Erboz (2017) and Dubey et al. (2022) and others.

Based on analysis (results from Table 4.2) the following conclusion of the research was formulated: all surveyed enterprises have knowledge of the Industry 4.0 concept. None of the respondents indicated a negative answer to the question Q1. Respondents are familiar with the pillars of Industry 4.0, but their knowledge of them varies. Among the responses from Q2_1 to Q2_9, the following pillars are well known: Big Data and Data Analytics (Q2_1), computer systems used for horizon and vertical integration (Q2_4), forms of additive manufacturing with 3D printing (Q2_8). The pillars of Industry 4.0 were rated highly, with respondents most often choosing 4 and 5 on the Likert scale (64% of responses for

Table 4.2 Structure of response shares in the RA_1 based on the questions: Q_1 and Q_2

RA_1: Knowledge of the concept of Industry 4.0 and its pillars					
Likert scale	*1*	*2*	*3*	*4*	*5*
Q_1. Do you have knowledge of the Industry 4.0 concept? (% of row)	0.0%	34.8%	29.2%	17.0%	19.0%
Q_2. Do you have knowledge of the pillars of Industry 4.0? Q2_1. Big Data and Analitycs (% of row)	2.0%	11.1%	22.9%	38.3%	25.7%
Q2_2. Autonomous robots/machine learning (ML) and other AI (% of row)	2.8%	51.0%	24.1%	15.0%	7.1%
Q2_3. Computer simulation/autonomous computer system/ autonomous system of process optimisation (% of row)	5.9%	18.2%	37.5%	27.3%	11.1%
Q2_4. Horizontal and Vertical System Integration/ computer systems/ICTs (% of row)	2.4%	7.9%	13.0%	48.2%	28.5%
Q2_5. IoT /IIoT (% of row)	4.0%	23.3%	28.5%	28.9%	15.4%
Q2_6. Cyber-Physical Systems (CPSs) and cybersecurity solutions (% of row)	9.1%	47.4%	26.9%	14.2%	2.4%
Q2_7. Cloud computing and cloud services (% of row)	4.0%	24.1%	25.7%	29.2%	17.0%
Q2_8. Additive manufacturing and 3D printing (% of row)	5.5%	23.3%	19.4%	30.8%	20.9%
Q2_9. Virtual and Augmented Reality/ HMI (Human–Machine Interface) (% of row)	5.9%	34.4%	27.3%	22.9%	9.5%

Source: own research

Q2_1, 76.7% for Q2_4 and 51.7% for Q2_8). The technological segment, formed from Big Data and Data Analytics, horizontal and vertical integration and additive manufacturing with 3D printing, is very well known by respondents, and their knowledge is practical, because the technologies of these pillars of Industry 4.0 are used in their companies.

The next step of the analysis was to determine if there is significant variation in the results of the questions (Q1 and Q2). The correlation between the questions responses was examined using Cumulative Link Mixed Models (CLMM) and One-way Repeated Ordinal Regression. These statistical techniques are applied to the analysis of ordinal outcome data, where observations are repeated measurements, usually with the same participants being evaluated under different circumstances or at different times. The independent variable is "Question", while "Respondents" are treated as a blocking variable (Table 4.3).

Table 4.3 Results of CLMM analysis using R computer program

Likelihood ratio tests of cumulative link models:

Formula	*Link*		*Threshold*		
model.null	Likert ~ 1		logit		
model	Likert ~ Question		logit		
No. parameters	*AIC*	*logLik*	*LR.stat*	*df*	*Pr(>Chisq)*
4	7496.3	-3744.2			
13	7174.0	-3574.0	340.28	9	< 2.2e-16 ***

Source: own research

The main effect of "Question" in the model was significant. At the post-analysis stage, the Tukey test (Table 4.4) was used for post hoc comparisons to evaluate the differences between pairs of questions.

As a result of the similarity analysis of the responses in the RA_1, groups from "a" to "e" were selected. Figure 4.2 presents these groups with median and mean. The first cluster is formed by Big Data& Data Analytics (Q2_1) and solutions in horizontal and vertical integrations (Q2_4). Respondents have good knowledge of computer systems for process support (Q2_4 in the group "a"). The second group (b) is formed by the cluster: Q2_5, Q2_7 and Q2_8. In this cluster are Internet of Things and Industrial Internet of Things, cloud computing, additive manufacturing with 3D printing. The last cluster that was formed from the lowest average ratings (group "de" and group "e"). Obtained ("de" and "e") cluster consists of two technological components: AI and CPPS solutions (Q2_2 and Q2_6).

The analysis shows that there is no significant statistical difference between the distribution of answers to questions Q2_4 and Q2_1, these answers created a group "a" in Fig. 4.1; the next groups are: Q2_8, Q2_7, Q2_5 (group "b"); Q_1 and Q2_3 (group "bc"); Q2_9 ("cd"); Q2_2 ("de"); and Q2_6 ("d"). Established groups have 2elements or 1-element. Tukey's test considers means. However, the distribution of data is equally important. For the Likert scale data, it is appropriate to analyse the distribution of the median, as well as the first and third quartiles. This analysis can yield different results when comparing responses to individual questions because it accounts for variability. Considering the ordinal

Table 4.4 Post hoc Tukey test comparisons using "R" computer program

contrast	*estimate*	*SE*	*df*	*z.ratio*	*p.value*
Q1 - Q2_1	-0.9665	0.162	Inf	-5.985	< .0001
Q1 - Q2_2	0.7801	0.162	Inf	4.83	0.0001
Q1 - Q2_3	-0.0717	0.159	Inf	-0.451	1
Q1 - Q2_4	-1.2745	0.162	Inf	-7.888	< .0001
Q1 - Q2_5	-0.2011	0.161	Inf	-1.249	0.9644
Q1 - Q2_6	1.0613	0.163	Inf	6.521	< .0001
Q1 - Q2_7	-0.2483	0.162	Inf	-1.533	0.8793
Q1 - Q2_8	-0.3952	0.165	Inf	-2.401	0.3253
Q1 - Q2_9	0.3539	0.162	Inf	2.187	0.4654
Q2_1 - Q2_2	1.7467	0.162	Inf	10.755	< .0001
Q2_1 - Q2_3	0.8949	0.158	Inf	5.668	< .0001
Q2_1 - Q2_4	-0.308	0.158	Inf	-1.953	0.6324
Q2_1 - Q2_5	0.7655	0.16	Inf	4.789	0.0001
Q2_1 - Q2_6	2.0279	0.164	Inf	12.374	< .0001
Q2_1 - Q2_7	0.7182	0.161	Inf	4.47	0.0003
Q2_1 - Q2_8	0.5714	0.163	Inf	3.509	0.0163
Q2_1 - Q2_9	1.3204	0.162	Inf	8.165	< .0001
Q2_2 - Q2_3	-0.8518	0.159	Inf	-5.366	< .0001
Q2_2 - Q2_4	-2.0547	0.163	Inf	-12.618	< .0001
Q2_2 - Q2_5	-0.9812	0.161	Inf	-6.091	< .0001
Q2_2 - Q2_6	0.2812	0.161	Inf	1.747	0.7684
Q2_2 - Q2_7	-1.0284	0.162	Inf	-6.345	< .0001
Q2_2 - Q2_8	-1.1753	0.165	Inf	-7.128	< .0001
Q2_2 - Q2_9	-0.4263	0.161	Inf	-2.647	0.1961
Q2_3 - Q2_4	-1.2029	0.158	Inf	-7.62	< .0001
Q2_3 - Q2_5	-0.1294	0.158	Inf	-0.821	0.9983
Q2_3 - Q2_6	1.133	0.16	Inf	7.079	< .0001
Q2_3 - Q2_7	-0.1766	0.159	Inf	-1.113	0.9835

(continued)

Table 4.4 (continued)

contrast	*estimate*	*SE*	*df*	*z.ratio*	*p.value*
Q2_3 - Q2_8	-0.3235	0.161	Inf	-2.006	0.5946
Q2_3 - Q2_9	0.4256	0.159	Inf	2.681	0.1811
Q2_4 - Q2_5	1.0735	0.16	Inf	6.721	< .0001
Q2_4 - Q2_6	2.3359	0.164	Inf	14.212	< .0001
Q2_4 - Q2_7	1.0262	0.16	Inf	6.394	< .0001
Q2_4 - Q2_8	0.8794	0.163	Inf	5.411	< .0001
Q2_4 - Q2_9	1.6284	0.162	Inf	10.057	< .0001
Q2_5 - Q2_6	1.2624	0.162	Inf	7.774	< .0001
Q2_5 - Q2_7	-0.0472	0.161	Inf	-0.294	1
Q2_5 - Q2_8	-0.1941	0.163	Inf	-1.188	0.9743
Q2_5 - Q2_9	0.555	0.161	Inf	3.447	0.0202
Q2_6 - Q2_7	-1.3097	0.163	Inf	-8.015	< .0001
Q2_6 - Q2_8	-1.4565	0.166	Inf	-8.763	< .0001
Q2_6 - Q2_9	-0.7075	0.162	Inf	-4.362	0.0005
Q2_7 - Q2_8	-0.1469	0.164	Inf	-0.894	0.9967
Q2_7 - Q2_9	0.6022	0.162	Inf	3.718	0.0077
Q2_8 - Q2_9	0.7491	0.165	Inf	4.55	0.0002

Source: own research

nature of Likert scale data, focusing on medians and quartiles provides a more accurate representation of the central tendency and variability of responses. The high median (median = 4) has the first group of answers („a"). This group presents the answers similarities for Big Data and Data Analytics (Q2_1), computer systems used for horizon and vertical integrations (Q2_4) from the group "a", and from the group "b" technologies of additive manufacturing including 3D printing (Q2_8). The lowest median (median = 2) is in two groups: group „de" (Q2_2) created by autonomous robots, machine learning (ML) and other solutions of AI, and group „d" (Q2_6): Cyber-Physical Systems (CPSs) and cybersecurity solutions. Median = 3 is in the other groups: computer simulation/ autonomous computer system/ autonomous process optimisation (Q2_

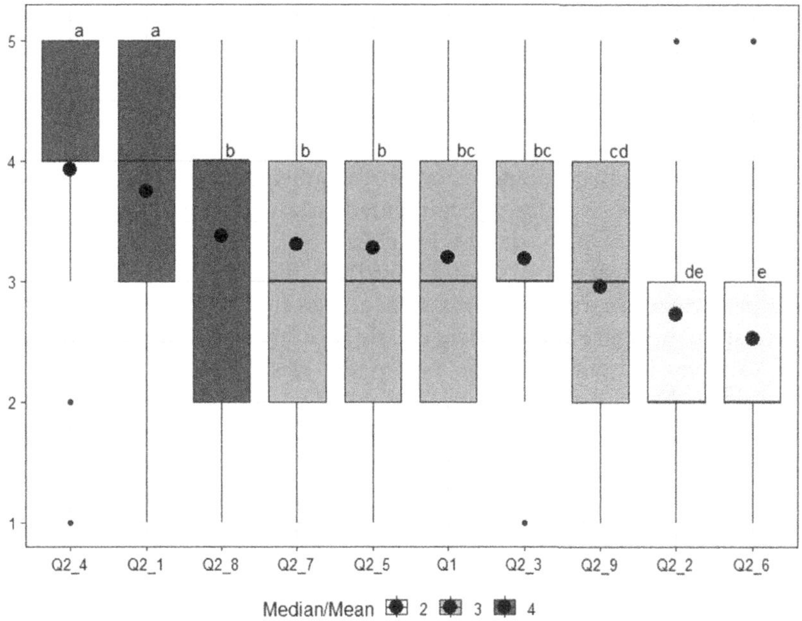

Fig. 4.2 Boxplot of Response Distribution and Median/Mean Comparison Across Questions with Tukey's HSD Test Results in the research area: RA_1. Source: own research

3), IoT and IIoT (Q2_5), cloud computing and cloud services (Q2_7) and Virtual and Augmented Reality/ HMI/Human–Machine Interface (Q2_9). Analysing and comparing quartiles, a similar distribution of responses was obtained for Q1, which refers to general knowledge about Industry 4.0, reinforced by three technologies: Q2_7, Q2_5 and Q2_9, representing IoT, Cloud Computing and Virtual and Augmented Reality. The level of knowledge is comparable for these three pillars. In the context of respondents' knowledge about specific pillars, 50% of respondents know very little about intelligent robots and cyber-physical systems (Q2_2 and Q2_6) and 50% of respondents have a high level of knowledge about computer systems supporting process integration of system modules (Q2_4).

The analysis performed shows the aggregate results of the questionnaire responses, but does not show the profile of the respondent (enterprises) that participated in the survey. Determining the profile of the respondent is important in this research because it allows us to show the interest of enterprises in Industry 4.0. Therefore, in the second stage of the analysis, a cluster analysis was conducted.

When trying to determine whether there are any similarities in the answers to individual questions for all respondents, it was found that the distribution of answers can be presented in a cluster analysis. Hopkins index of 0.536, which should be close to 1 (Hopkins index is not below 0.5 so cluster analysis is allowed). A hierarchical cluster analysis was performed for the questions. When assessing the fit of clusters, the correlation between the cophenetic distance and the distance between the variables is examined. The best result was given by the complete method with Manhattan distance (Manhattan distances were used due to the nature of the data—on the Likert scale), where the correlation was 0.62. This indicates a moderate fit to the data and allows for further analysis. The results of the analysis are presented in the dendrograms. Figure 4.3 presents such results for the RA_1. In this figure, a division into 7 clusters is used. This is due to the study of individual indices: mean silhouette score, dunn index and davies bouldin index. For this division, these indices gave the best values: 0.2222825, 1.358817 and 0.5371881. Figure 4.4 presents the results of the Manhattan analysis. Based on the distance matrix between questions with Manhattan distance, strongly related questions were obtained (analysis of the smallest distances).

Based on the dendrogram analysis (Fig. 4.3) following average linkage between groups was found: Q2_3 and Q2_5 (this cluster is formed from: solutions of computer simulation/autonomous computer systems/autonomous process optimisation and IoT /IIoT); Q2_1 and Q2_4 (Big Data &Big Data Analytics and Horizontal and Vertical System Integration), Q1and Q2_7 (concept of Industry 4.0 and cloud computing with cloud services). In the analysis performed, no strong connection was obtained between respondents' general knowledge of the Industry 4.0 concept and its pillars. Such a connection was only related to the cloud computing, so it can be presumed that for respondents Industry 4.0 is associated with the cloud (the first association of respondents).

In the Fig. 4.4, the smallest distance is between Q2_3 and Q2_5 (where: Q2_3 is computer systems and computer simulation and Q2_5 is IoT) and between Q2_4 and Q2_1 (where: Q2_4 is horizontal and vertical integration and Q2_1 is Big Data and Data Analytics) and the largest between Q2_4 and Q2_6 (where Q2_4 is horizontal and

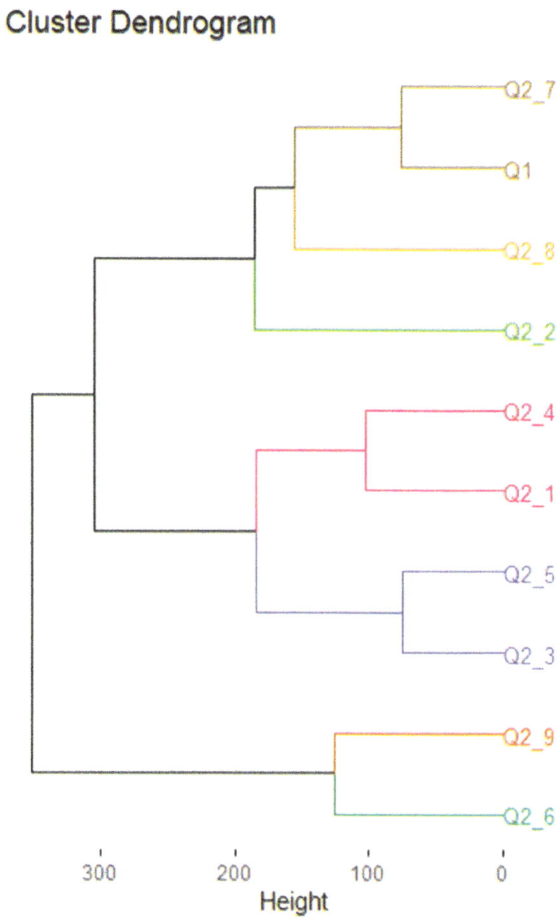

Fig. 4.3 Cluster dendrogram of response similarity in the research area: RA_1. Source: own research

vertical integration and Q2_6 is CPS). For respondents, the least obvious according to their knowledge of Industry 4.0 is technology integration in CPSs.

Both analyses (Figs. 4.3 and 4.4) confirmed that Q2_4 and Q2_1 are the key clusters initiating the smart transformation. On the other hand, it is interesting to note that the two clusters Q2_3 and Q2_5 (computer

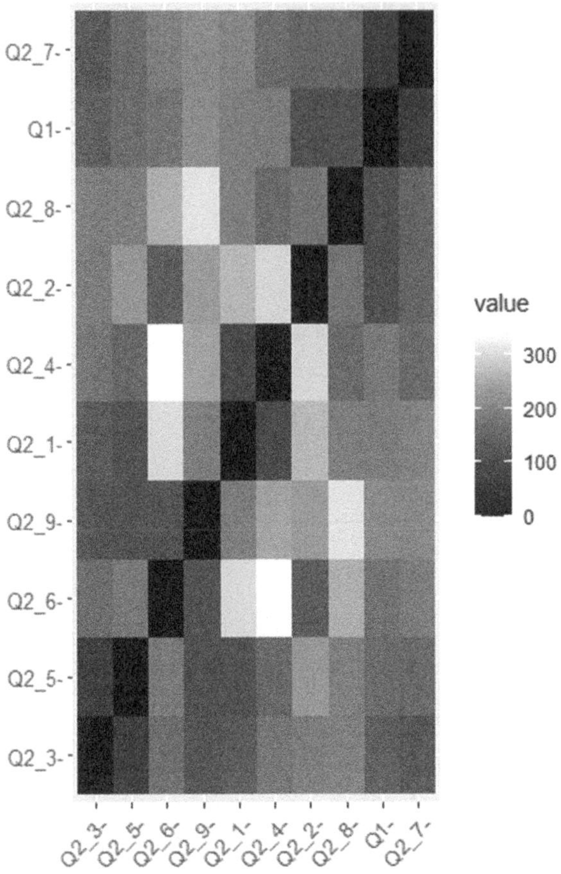

Fig. 4.4 Distance matrix between questions with Manhattan distance in the research area: RA_1. Source: own research

systems/computer simulation/computerised process support systems and IoT) had the most similar responses from the same respondents, which may indicate a strong connection in the transformation of these two pillars in progress.

In partitioning clustering (clustering methods used to classify observations in a data set into multiple groups based on their similarity) to determine the number of clusters to be generated can be used: K-medoids clustering or PAM (Partitioning Around Medoids, Kaufman & Rousseeuw, 1990), in which each cluster is represented by one of the objects in the cluster. PAM is less sensitive to lag values compared to k-means.

Before conducting the analysis using the non-hierarchical method, it was examined, using the Hopkins criterion, whether clusters could be distinguished. The index of 0.756[1] was obtained, which shows that clusters can be extracted. Next, it was examined using PAM how many clusters can be built. It was determined that two clusters could be extracted. Figure 4.5 presents the received two clusters based on the results of statistical analysis (Table 4.5).

Cluster 1 (a larger cluster) includes respondents with scores of 3 and below. This cluster groups 174 companies. Cluster 2 (a smaller cluster) brings together respondents giving the highest ratings on the Likert scale. The cluster groups 71 companies. Each cluster has its leader. The leaders are described in the Tables 4.6 and 4.7. In cluster 1, the respondent (leader) had the number 241, and in cluster 2 number 221.

Based on the discrepancy analysis (Fig. 4.6), a strong discrepancy (for 4 and 5 ranks) was obtained in Q2_2 (knowledge of AI—autonomous robots—as a pillar of Industry 4.0) and Q2_9 (knowledge of virtual and augmented reality). These two areas are the most difficult to implement the pillars of I 4.0 at the current stage of implementation of the concept. Knowledge of these pillars is incomplete due to the wide range of technological solutions embedded in AI and augmented reality. It is difficult for manufacturing companies to make decisions on their own about the direction of implementation of these pillars, which is why they need the support of the IT industry (external companies)—consulting

[1] Hopkins statistic was calculated in R using the package {factoextra}

An example: get_clust_tendency(ques, 2, graph = TRUE, gradient = list(low = "red", mid = "white", high = "blue"), seed = 123)$hopkins_stat [1] 0.5359647.

get_clust_tendency(df, 2, graph = TRUE, gradient = list(low = "red", mid = "white", high = "blue"), seed = 123).

$hopkins_stat [1] 0.7560813.

Hopkins statistic: If the value of Hopkins statistic is close to 1 (far above 0.5), then we can conclude that the dataset is significantly clusterable.

VAT (Visual Assessment of cluster Tendency): The VAT detects the clustering tendency in a visual form by counting the number of square shaped dark (or coloured) blocks along the diagonal in a VAT image.

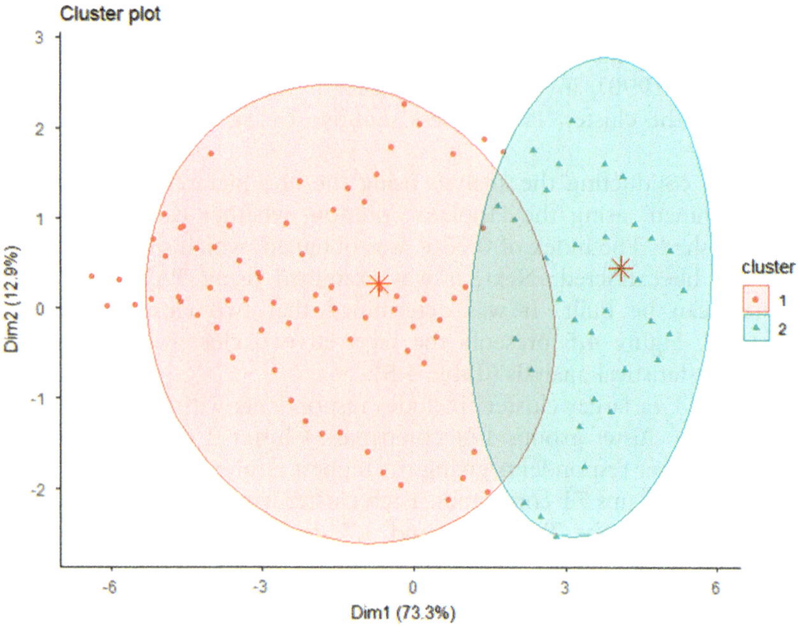

Fig. 4.5 K-medoids cluster analysis in the research area: RA_1. Source: own research

Table 4.5 Score Method Clusters

APN 0.05394254 pam	*2*
AD 4.83018258 pam	6
ADM 0.29469338 k-means	3
FOM 0.58790671 pam	6

Table 4.6 Answers of respondent 214 (medoid of the cluster analysis)

Q_1	*Q2_1*	*Q2_2*	*Q2_3*	*Q2_4*	*Q2_5*	*Q2_6*	*Q2_7*	*Q2_8*	*Q2_9*
3	3	3	3	4	3	2	3	3	3

Source: own research

Table 4.7 Answers of respondent 221 (medoid of the cluster analysis)

Q_1	*Q2_1*	*Q2_2*	*Q2_3*	*Q2_4*	*Q2_5*	*Q2_6*	*Q2_7*	*Q2_8*	*Q2_9*
5	5	4	4	5	5	4	5	5	4

Source: own research

services during project management stage. AI and augmented reality are new technologies (pillars of smart transformation for enterprises in the steel market) that may yet surprise us. Artificial intelligence (AI) continues to develop at a rapid pace, and its impact on our lives is becoming more and more apparent. So, are we heading towards so-called general artificial intelligence (AGI), which aspires to reach a level of intelligence comparable to humans?

The vertical and horizontal analysis is summarized in a scatter plot (Fig. 4.7). This graph summarises the pairwise correlation between all questions and all answers (pairwise correlation). The narrower the ellipse, the higher the Pearson linear correlation. The following correlations was established: Ql and Q2_7, Qland Q2_3, Q2_1and Q2_5, Q2_1 and Q2_9, Q2_2 and Q2_7, Q2_3 and Q2_5, Q2_3 and Q2-6, Q2_3 and Q2_4.

RESEARCH AREA: RA_2

This area of research was used to determine the impact of the technological pillars of Industry 4.0 on the changes in the Polish steel industry. Two questions were analysed (Q3 and Q4). The question Q3 was about the impact of the Industry 4.0 pillars on industrial transformation of the steel enterprises: Q3: What is (in your opinion) the impact of the concept of Industry 4.0 and its technological pillars on changes in your enterprise?). Respondents had 5 proposed answers from Q3_1 to Q3_9 according to the Likert scale from 1 to 5, where: very low = 1 point, low = 2 points, average = 3 points, high = 4 points, very high = 5 points. In this question were analysed general pillars of the Industry 4.0 concept. In the next question (Q4) was analysed particular solutions of technological pillars. The question was: Q4: How important are technological solutions of Industry 4.0 for your enterprise in smart transformation? Respondents had 12 proposed answers from Q4_1 to Q4_12 described by the Likert scale from 1 to 5 (Table 4.8).

Based on the answers to question Q3, the following conclusion was made: respondents considered all pillars of Industry 4.0 to have a strong

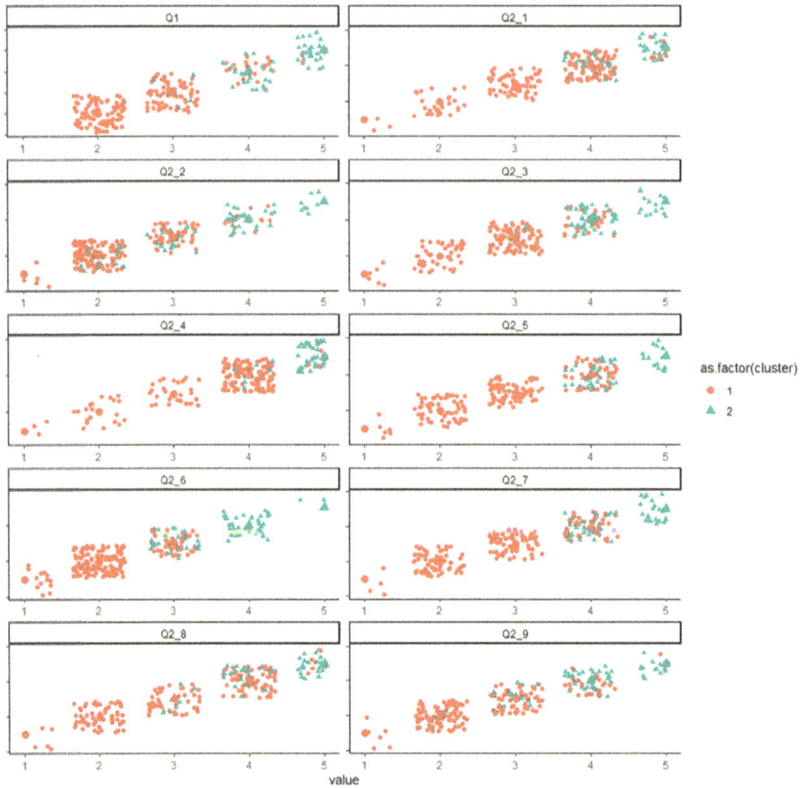

Fig. 4.6 Discrepancy analysis in the research area: RA_1. Source: own research

impact on the transformation of the steel industry. In the detailed analysis (Q4), high shares of ranks (4 and 5) were for the following technological solutions: production automation with machine islands (Q4_1) and robots (Q4_3), intelligent sensors and control systems (Q4_7), solution based on Big Data and Data Analytics (Q4_10), computer systems helping in horizontal and vertical integration of processes (Q4_9) and blockchain technologies (Q4_12). Based on the respondents' answers to Q4, the next conclusion was drawn: the industrial transformation of steel companies is based on automation and robotisation of production and logistics with support of processes by computer systems that improve decision-making and communication. New technologies are important

Fig. 4.7 Scatter plot of horizontal and vertical analysis in the research area: RA_1. Source: own research

Table 4.8 Structure of response shares in the RA_2 based on the questions: Q_3 and Q_4

Impact of Industry 4.0 on industrial transfromation of steel enterprises						
Question	*no res*	*1*	*2*	*3*	*4*	*5*
Q3_1. Big Data and Data Analytics	2.0%	0.0%	5.1%	13.0%	34.0%	45.8%
Q3_2. Intelligent and autonomous robots, machine learning and other AI solutions	2.8%	1.6%	8.7%	28.5%	42.7%	15.8%
Q3_3. Computer simulation systems and visualization systems/computer modeling/computer prediction	5.9%	0.8%	4.3%	24.1%	46.2%	18.6%
Q3_4. Computer integrated systems and horizontal and vertical integration	2.4%	0.0%	2.0%	9.5%	47.0%	39.1%
Q3_5. IoT and IIoT	4.0%	0.0%	3.6%	16.2%	41.1%	35.2%
Q3_6. Cyber Physical Production Systems and cybersecurity tools	9.1%	0.4%	0.8%	14.2%	47.0%	28.5%
Q3_7. Cloud computing and cloud services	4.0%	0.0%	11.1%	25.7%	40.7%	18.6%
Q3_8. Additive manufacturing including 3D printing and others	5.5%	0.0%	2.0%	13.4%	42.7%	36.4%
Q3_9. Virtual and augmented reality	5.9%	0.0%	2.4%	8.3%	39.9%	43.5%
Q4_1. Automation of production (from single machine to islands of machines)	0.0%	0.0%	6.3%	27.3%	41.9%	24.5%
Q4_2. Full automatization of key operations in manufacturing process	0.0%	0.0%	9.1%	29.2%	43.1%	18.6%
Q4_3. Industrial robots and cobots in processes	0.0%	0.0%	4.3%	19.4%	43.1%	33.2%
Q4_4. Full automatization of storage process	0.0%	0.0%	7.5%	27.7%	37.2%	27.7%
Q4_5. Advanced changes towards smart manufacturing	0.0%	0.0%	8.7%	29.2%	41.1%	20.9%
Q4_6. Mobile technologies in consumer services including e-invoice system, EDI and others	0.0%	0.0%	9.5%	26.9%	37.5%	26.1%
Q4_7. Intelligent sensors, real data analysis, autonomous system of decision making, predictive maintenance and others	0.0%	0.0%	2.8%	13.4%	41.1%	42.7%
Q4_8. Computers systems of processes support (e.g. CAD, MES, ERP, others)	0.0%	0.0%	4.3%	21.7%	46.2%	27.7%
Q4_9. Intelligent value chain and network cooperation	0.0%	0.0%	2.0%	9.9%	46.2%	41.9%
Q4_10. Databases and open data, real time of data analysis	0.0%	0.0%	7.1%	13.0%	36.0%	43.9%

(continued)

Table 4.8 (continued)

Impact of Industry 4.0 on industrial transfromation of steel enterprises						
Question	*no res*	*1*	*2*	*3*	*4*	*5*
Q4_11. Cloud computing and cloud services, IIoT	0.0%	0.0%	12.6%	22.1%	41.9%	23.3%
Q4_12. Technology blockchain	0.0%	0.0%	2.8%	10.3%	45.8%	41.1%

Source: own research

both for the development of steel enterprises, and for better cooperation in supply chains. Steelmakers are strongly connected with both suppliers of materials and consumers of steel products. Respondents are aware of the transformation direction of smart steel manufacturing and smart steel value chains. Figure 4.8 presents the analysis of response similarity in the RA_2/Q3 and Q4. In this analysis, responses with a value of "0″ (no response) were removed. Groups separated based on the similarity of responses are one-element, two-element and three-element. Median = 4 was in all separated groups. In assessing the impact of Industry 4.0 technologies on the transformation of the steel industry, the groups with the highest ratings formed the following technologies: Virtual and Augmented Reality/ HMI (Human–Machine Interface) (group „a" formed from Q3_9), Horizontal and Vertical System Integrations, solutions building intelligent value chain and smart manufacturing in CPPS and Blockchain (group „ab" formed from Q4_9, Q3_4, Q4_12), autonomous system of decisions, predictive maintenance (PM), Big Data and Data Analytics with databases and computer system of decision-making, technologies of additive manufacturing including 3D printing, IoT and IIoT and other solutions of CPPS including cybersecurity (group "abc" formed from Q4_7, Q3_1, Q3_8, Q4_10, Q3_6 and Q3_5). In the process of industrial transformation of steel enterprises towards Industry 4.0, the important place have the industrial robots (cobots) (group "bcd" formed from Q4_3). Analysis of questions Q3 and Q4 together as an analysis of the impact from the macro level to the micro level separated segments with little differentiation and a lot of clusters were identified.

Performed Hopkins Test of 0.522 (above 0.5) for questions as observations (horizontal analysis) allows for cluster analysis. A hierarchical cluster analysis (HCA) was used to select the cluster model. In an analogous

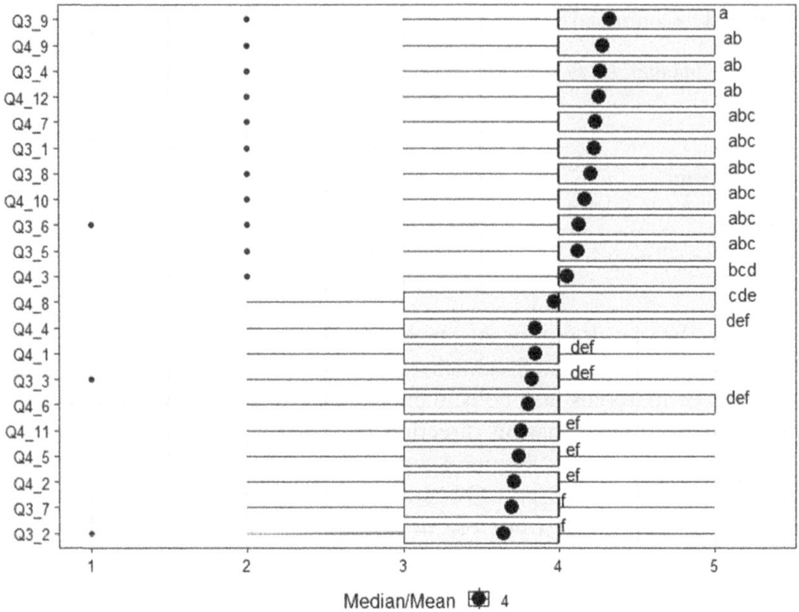

Fig. 4.8 Boxplot of Response Distribution and Median/Mean Comparison Across Questions with Tukey's HSD Test Results in the research area: RA_2. Source: own research

way to the first area, a dendrogram was used to visualise the dendrogram (Fig. 4.9).

Based on the dendrogram (Fig. 4.9) following average linkage between groups were found: Q3_1 and Q4_10 (this cluster is formed from: Big Data& Data Analytics and Database and open data); Q3_4 and Q4_9 (computer integrated systems with horizontal and vertical integration is linked to intelligent value chain and networks); Q3_5 and Q4_7 (IoT and IIoT linked to intelligent sensors, real data analysis, autonomous systems of decision-making, predictive maintenance and others); Q3_9 and Q4_12 (Virtual and Augmented Reality linked to technology blockchain); Q3_7, Q4_6, Q4_11 (cloud computing and cloud services, mobile technologies in consumer services, e-invoice, EDI, QR and others, IoT and IIoT); Q4_2, Q3_2, Q4_5, Q4_1, Q4_4 (processes automatisation, intelligent and autonomous robots, machine learning and other

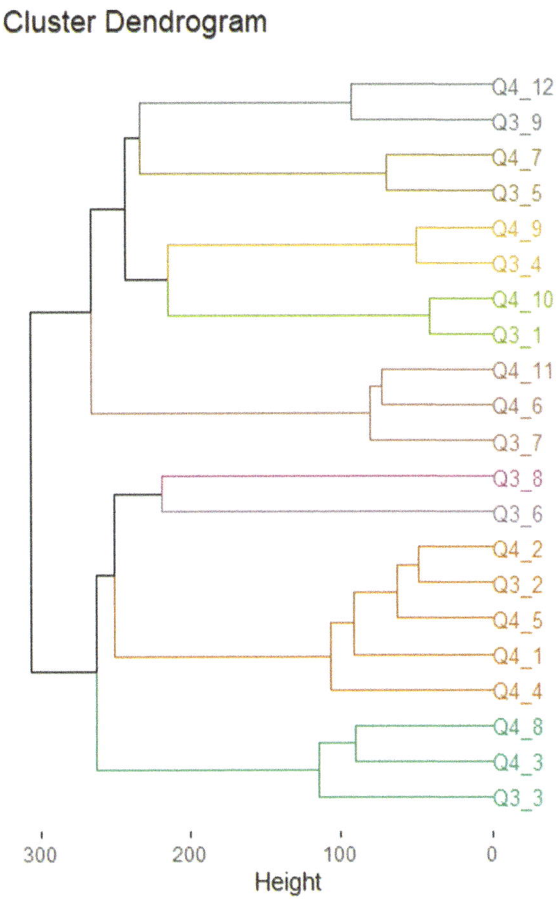

Fig. 4.9 Cluster dendrogram of response similarity in the research area: RA_2. Source: own research

AI solutions, smart manufacturing with automation of manufacturing and storage processes); Q4_8, Q4_3, Q3_3 (computers systems: CAD, MES, ERP, others, robots and cobots, computer simulation systems and processes visualisation, computer modelling and computer prediction). Q3_6: CPPS and Q3_8: Additive manufacturing including 3D printing are not part of any clusters. The answers to this question deviate the most from the others.

The in-depth cluster analysis performed shows that they are strongly drawn (Fig. 4.10) the cluster: Q4_2, Q3_2, Q4_5, Q4_1, Q4_4. This is the largest cluster formed from a set of following technologies and changes towards full automatisation of manufacturing and storage processes with intelligent and autonomous robots, machine learning (ML) and other AI solutions. In this cluster, there is a strong relationship in the cluster formed from Q4_2, Q3_2, Q4_5 (high automatisation of manufacturing with robots and other solutions towards smart manufacturing). The darkest colour indicates the smallest distance for the two clusters: the first cluster is formed from Big Data& Data Analytics with databases (Q3_1 and Q4_10), the second cluster is formed from computer integrated systems and Horizontal and Vertical System Integration and changes towards intelligent value chain of the steel industry (Q3_4 and Q4_9).

The next stage of the analysis involved determining the profile of the respondent—a representative of a steel company—providing answers to the directions of technological change in his company. In order to check the clustering of the observations, the Hopkins index was calculated, which was significantly higher than the 0.5 level, at 0.698, which entitled the cluster analysis of the respondent's profile. At this stage, the Hidden Markov Model (HMM) analysis was used to select the cluster model. The HMM model defines a set of states that a system can be in. The number of these states is the key parameter of the model. Two states were established for the application of technologies in steel enterprises by respondents with knowledge of Industry 4.0:

- State 1: Respondents who have basic knowledge and are committed to applying selected Industry 4.0 technologies.
- State 2: Respondents who have advanced knowledge and are applying a wide variety of technologies at the stage of building smart manufacturing in their enterprises.

It was decided not to include more states because respondents with low knowledge and low engagement in smart transformation of steel enterprises in Poland were in the minority.

Using the AIC (Akaike Information Criterion) and BIC (Bayesian Information Criterion) and the model fit criterion, which is a measure

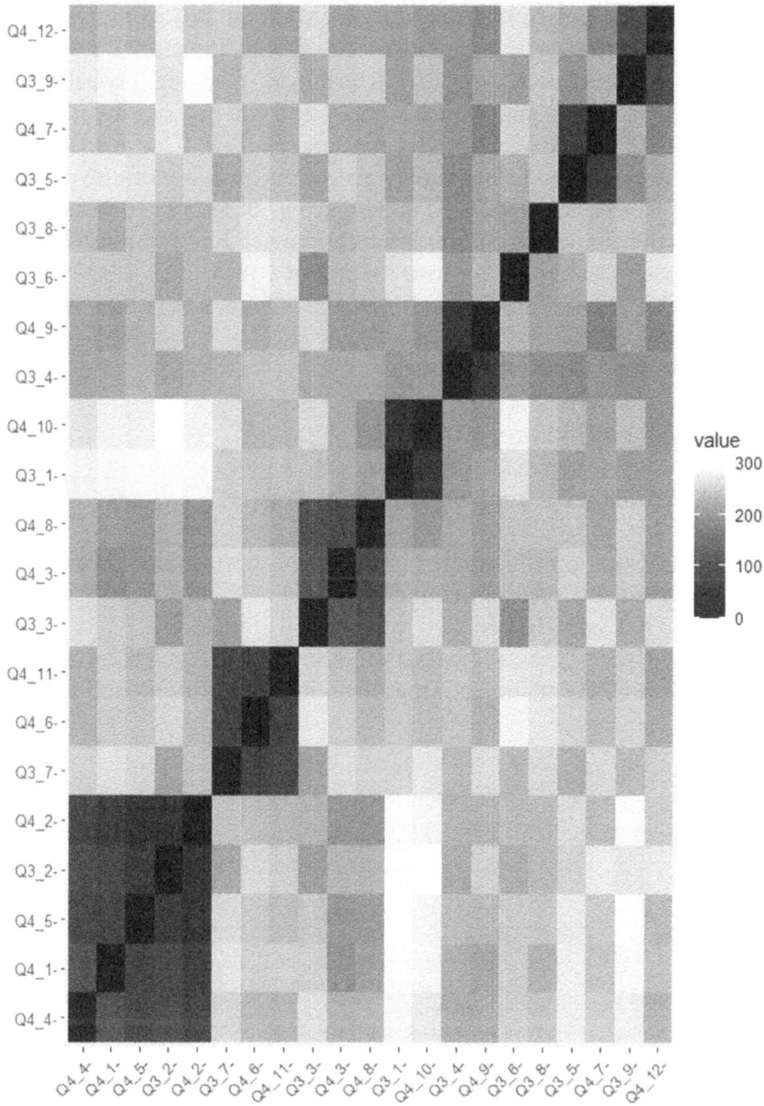

Fig. 4.10 Distance matrix between questions with Manhattan distance in the research area: RA_2. Source: own analysis

of model quality, where smaller values of AIC and BIC indicate a better fit of the model to the data, the model recorded in Table 4.9 was selected.

Based on the analysis, it was determined that the probability of remaining in state 1 is approximately 56%, and the probability of moving from state one to the second is approximately 44%. The probability of remaining in the second state is 69% and the probability of moving from the second to the first state is 31%. The group of leaders consists of 148 companies that indicated important technologies for the transformation of this industry (at least 75% chose a rating of 4 or 5): Q3_1, Q3_2, Q3_4, Q3_5, Q3_6, Q3_8, Q3_9, Q4_1, Q4_2, Q4_3, Q4_4, Q4_5, Q4_7, Q4_8, Q4_9, Q4_10, Q4_12.

The full list of these technologies:

according to Industry 4.0 pillars
Q3_1. Big Data& Data Analytics
Q3_2. Intelligent and autonomous robots, ML and other AI solutions
Q3_4. Computer-integrated systems in supporting processes and horizontal and vertical integration
Q3_5. IoT and IIoT
Q3_6. Cyber-Physical Production Systems with cybersecurity tools
Q3_8. Additive manufacturing, including 3D printing and others

Table 4.9 AI and BIC analysis using "R" computer program

```
print(transition_matrix)
[[1]]
Model of type multinomial (identity), formula: ~ 1
< environment: 0 × 0000023113fe21a0 >
Coefficients:
[1] 0.5576923 0.4423077
[[2]]
Model of type multinomial (identity), formula: ~ 1
< environment: 0 × 0000023113fe21a0 >
Coefficients:
[1] 0.3108108 0.6891892
> logLik(fit_model)
'log Lik.' -5979.746 (df = 155)
> AIC(fit_model)
[1] 12,269.49
> BIC(fit_model)
[1] 12,817.17
```

Source: "R" program stat. analysis

Q3_9. Virtual and Augmented Reality
according to changes in enterprises:
Q4_1. Automation of production (the first step: single high machines, the next step: islands of high machines)
Q4_2. Full (high) automatisation of key operations in manufacturing processes
Q4_4. Full (high) automatisation of storage processes
Q4_5. Advanced changes towards smart manufacturing
Q4_7. Intelligent sensors, real data analysis, autonomous systems of decision-making, predictive maintenance and others
Q4_8. Computers systems of processes (e.g. CAD, MES, ERP, others)
Q4_9. Intelligent value chain with strong network cooperation
Q4_10. Access to databases in real time of data analysis and open data
Q4_12. Technology blockchain

Presentation of the results of the analysis in Fig. 4.11. Description for the Fig. 4.11: dash is the median, dots are outliers. The difference between the states is greatest in: Q3_2, Q3_6, Q3_8, Q4_1, Q4_2, Q4_4, Q4_5, Q4_8.

Research Area: RA_3

The third area of research (RA_3) was about main applications/ in-depth implementations (directions of change) of technological changes based on pillars of Industry 4.0. This research area was formed from two questions: Q5: What changes are realised in your enterprise in industrial transformation to Industry 4.0? Q6: What results of change do you expect from new technologies (kind of expectations)? The shares of responses according to the Likert scale are shown in Table 4.10.

Based on the structure of the responses (Table 4.10), the values (total shares of responses) for ranks 4 and 5 were analysed. For question Q5, the highest values were for the following implementations: integration of computer and communication systems in enterprises and in supply chains (intra-and extra-groups) (73.5%), access to Big Data and Data Analytics with intelligent decision-making systems in enterprises (69.2%) and access to blockchain technology (69.2%). The high shares of responses for the listed solutions formed the basis for the conclusion: the transformation of steel companies is based on ICTs technologies that integrate processes

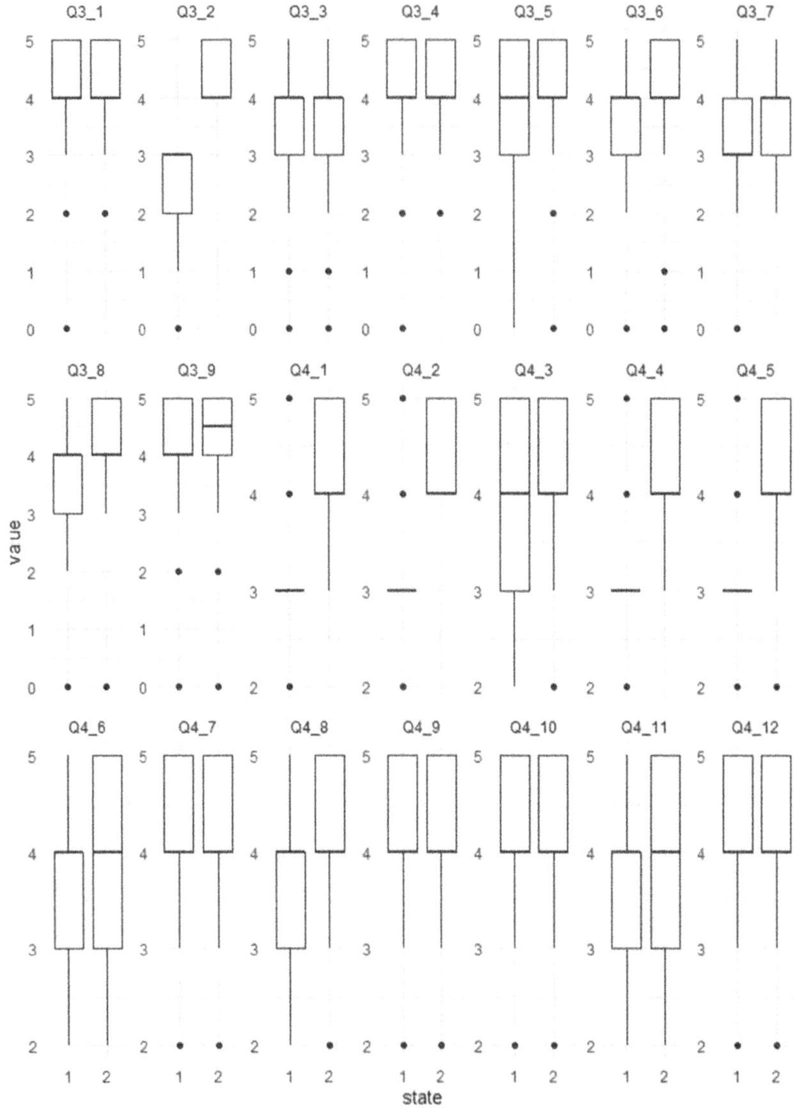

Fig. 4.11 Boxplots for observed states in HMM. Source: own research

Table 4.10 Structure of response shares in the RA_3 based on the questions: Q5 and Q6

Diagnosis of changes in steel industrial transformation					
Question	*1*	*2*	*3*	*4*	*5*
Q5_1. Automation of the manufacturing process (advanced machines, ICTs)	6.7%	11.1%	27.3%	41.5%	13.4%
Q5_2.Automation of the manufacturing process with M2M	4.0%	13.8%	32.4%	40.7%	9.1%
Q5_3. Intelligent robots/robot assembly/ autonomous vehicles	4.0%	9.5%	20.9%	48.2%	17.4%
Q5_4. Automation of warehouses and logistics processes	4.7%	9.5%	28.9%	41.1%	15.8%
Q5_5. Automation of manufacturing processes with multi-tasking process support computer systems	4.3%	15.4%	28.1%	41.1%	11.1%
Q5_6. Development of mobile technology in the customer service process	3.6%	9.5%	26.9%	41.5%	18.6%
Q5_7. Installation of sensors generating machine performance data	2.4%	11.9%	21.3%	38.3%	26.1%
Q5_8. Development and compatibility of IT systems (CAx, MES, etc.)	4.0%	9.9%	24.1%	44.3%	17.8%
Q5_9. Development and compatibility of IT systems within the framework of network and chain integration (intra- and extra-group)	4.3%	5.5%	16.6%	46.6%	26.9%
Q5_10. Big data, transfer data in real time, optimization of processes by ICTs	2.8%	11.5%	16.6%	35.2%	34.0%
Q5_11. Cloud computing and clud services/ IoT and IIoT	2.4%	15.4%	24.5%	41.9%	15.8%
Q5_12. Blockchain technology	2.4%	10.3%	18.2%	35.6%	33.6%
Q6_1. Process optimisation and productivity	0.8%	7.1%	25.3%	39.5%	27.3%
Q6_2. Real-time analysis and autonomous computer control systems	0.4%	5.5%	23.7%	38.3%	32.0%
Q6_3.Capacity/processes optimisation/ autonomous systems for decision-making	0.0%	5.9%	24.1%	38.3%	31.6%
Q6_4. Flexibility of operations/processes/ supply chains	0.0%	6.7%	24.1%	34.0%	35.2%
Q6_5. Labor cost reduction and reorganization	0.0%	2.0%	13.0%	38.7%	46.2%
Q6_6. Product development (quality, product accuracy, repeatability, etc.)	0.0%	1.2%	6.3%	46.6%	45.8%
Q6_7. Reduction of operating costs (FMS, autonomous technologies)	0.0%	5.9%	31.6%	43.5%	19.0%
Q6_8.Product presonalisation, castomisation, customer services	0.0%	6.3%	31.2%	43.5%	19.0%

(continued)

Table 4.10 (continued)

Diagnosis of changes in steel industrial transformation					
Question	*1*	*2*	*3*	*4*	*5*
Q6_9. Higher market value of the enterprise	0.0%	5.9%	30.4%	44.7%	19.0%
Q6_10. Flexible distribution systems/speed and quality of deliveries	0.0%	0.4%	5.9%	49.4%	44.3%
Q6_11.Supply chain management and processes integration	0.0%	0.0%	5.5%	53.0%	41.5%

Source: own research

in companies and supply chains. Automation of operations and robotisation of processes (Q5_1, Q5_3) with support from computerized systems (CAD, ERP, MES, etc.) are important areas of process improvement in companies (Q5_8). Improvements in market communication and the development of electronic customer service processes are also important areas of changes in personalisation (Q5_6).

In the analysis of the expected results of the implemented solutions (Q6), three responses exceeded the shares of 90% of the total responses, and they were: (Q6_11) supply chain management, (Q6_10) flexible distribution, speed and quality of deliveries (93.7%), product development and personalisation (Q6_6). Based on the structure of these responses, a conclusion was reached: the technologies of Industry 4.0 should contribute to the development of steel products and the construction of flexible supply chains. Figure 4.12 presents vertical analysis response similarity in the in RA_3 based on the questions: Q5 and Q6.

Respondents highly rated the improvements that make up the group "a". This group consists of the following answers: Q6_10, Q6_6, Q6_ 11, Q6_5. Based on this group, the following conclusion was made: technological improvements of Industry 4.0 are expected to improve supply chains (Q6_10 and Q6_11) and steel product development (Q6_6), as well as (Q6_5) reduce manufacturing costs. This conclusion is in line with the conclusion drawn from the distribution of responses included in Table 4.9.

Figure 4.13 presents a cluster dendrogram. In this figure the following clusters are presented: Q6_11, Q6_10, Q6_6; and Q6_1, Q6_2, Q6_ 3, Q6_4, Q6_7, Q6_8, Q6_9. The first cluster form changes towards: supply chain management and processes integration (Q6-11), flexible

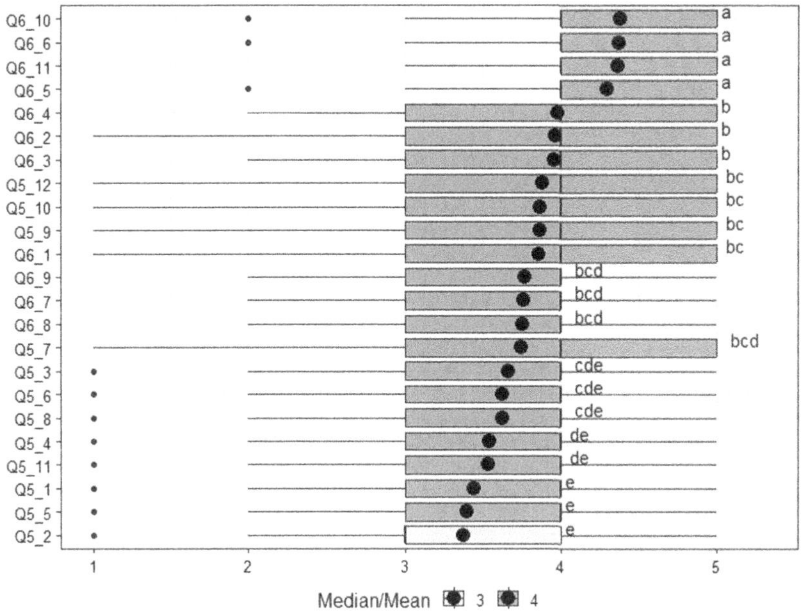

Fig. 4.12 Boxplot of Response Distribution and Median/Mean Comparison Across Questions with Tukey's HSD Test Results in the research area: RA_3. Source: own research

distribution systems and speed deliveries (Q6_10), product development (quality, product accuracy, repeatability, etc.) and personalisation (Q6_6). The second cluster is larger than the first, it was formed by changes towards process optimisation and productivity (Q6_1), real-time analysis and autonomous computer control systems (Q6_2), capacity by processes optimisation and autonomous systems for decision-making (Q6_3), flexibility of operations and processes in enterprises and in supply chains (Q6_4), reduction of operating costs by FMS and agile (Q6_7), product presonalisation, castomisation, customer services (Q6_8) and higher market value of the enterprise (Q6_9).

Based on Hopkins stat. 0.551174 and cor(dist.man.question3, res.coph3) 0.8337212 (distances are well correlated with cophenic distance) it was possible to form clusters, which is shown in Fig. 4.14.

Based on the discrepancy analysis (Hopkins_stat. 0.6055), a strong discrepancy (for 4 and 5 ranks) was obtained in: Q5_1, Q5_2, Q6_1, Q6_

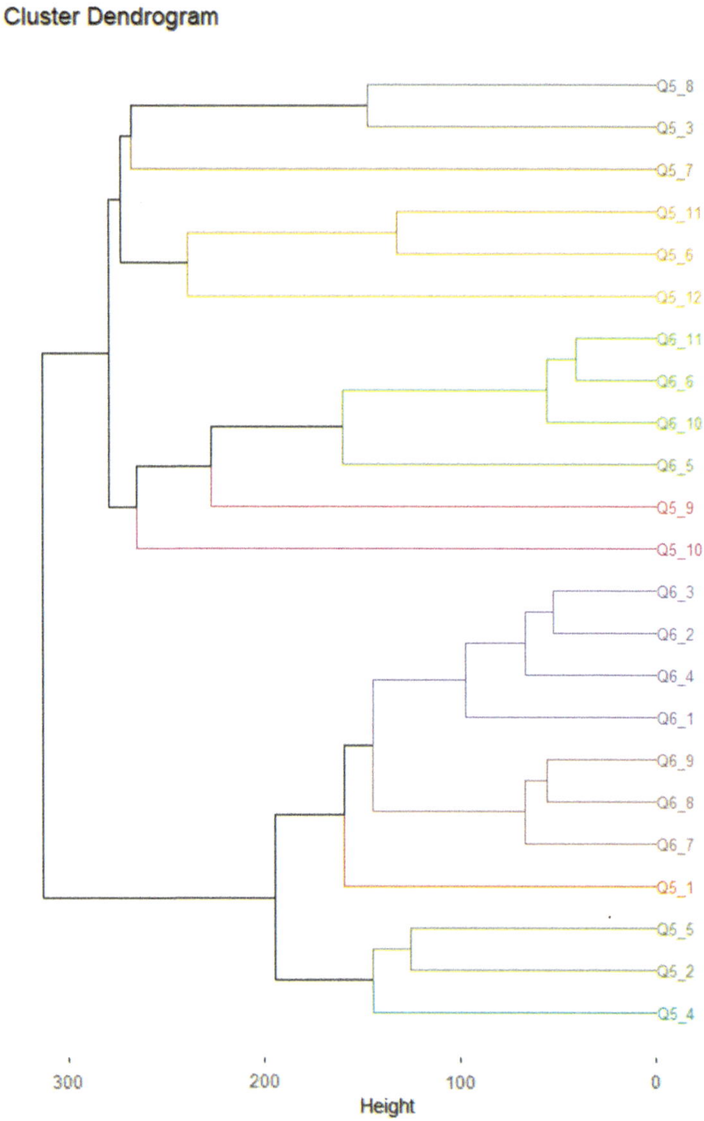

Fig. 4.13 Cluster dendrogram of response similarity in the research area: RA_3. Source: own research

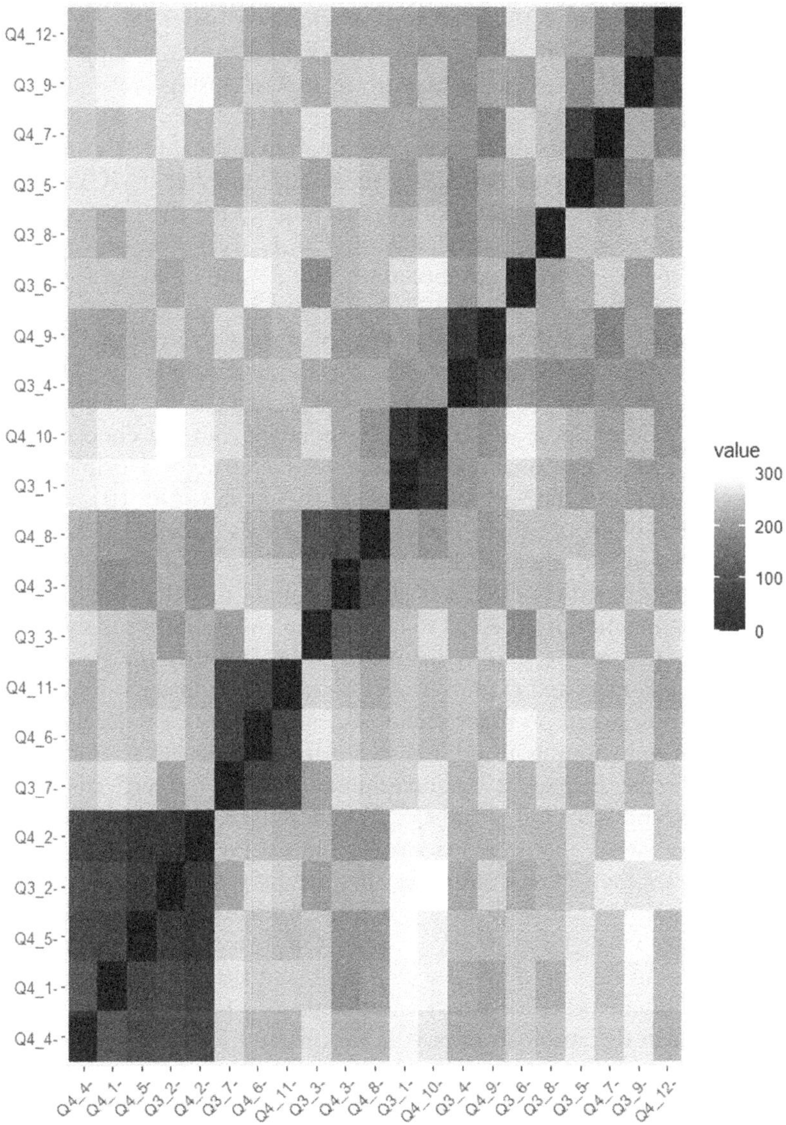

Fig. 4.14 Distance matrix between questions with Manhattan distance in the research area: RA_3. Source: own research

2, Q6_3, Q6_4 (Fig. 4.15). The "green" cluster is better than the "red" cluster (Fig. 4.15). For answer Q6_11 (changes towards supply chain management and processes integration) images of clusters are similar (for ranks 3, 4, 5).

Further investments in process automation with M2M (machine-to-machine) components, alongside substantial funding in ICTS (Q5_1, Q5_2). These investments are, according to the respondents, contributing to: increased productivity (Q6_1), real-time process data analysis (Q6_2), process optimisation (Q6_3) and enhanced process flexibility (Q6_4).

The analysis also indicates that there are areas where respondents did not select the lowest-ranking options. These areas include Q6_11, Q6_10, Q6_9, Q6_8, Q6_7, Q6_6, Q6_5, Q6_4 and Q6_3. Among these, a particularly interesting area is where respondents did not choose either 1 or 2. This area comprises two outcomes of the changes: Flexible distribution systems and the speed and quality of supplies (Q6_10), and the improvement of supply chain management systems (Q6_11). This trend was observed in both the first cluster (red) and the second cluster (green). Respondents, regardless of cluster affiliation, indicated that the ongoing smart transformation is expected to lead to improvements in supply chains within the steel sector. This is justified by the fact that the market for suppliers, customers and manufacturers in the steel industry is global and requires advancements in smart technologies. Technological development ensures not only optimisation but also enhances resilience.

Based on the analysis, it was determined that the "green" cluster (in terms of selections for Q5_1 and Q5_2) indicated that the level of process automation within enterprises is already high or very high (Fig. 4.15). This suggests that these companies may be moving to the next stage of process automation, with strong support from AI. Unfortunately, the "red" cluster (Fig. 4.15) in these areas (Q5_1 and Q5_2) shows that there are companies in the steel sector in Poland that are only beginning to lay the foundations for smart transformation, focusing primarily on investments in the automation of core processes.

Research Area: RA_4

In the fourth research area (RA_4), it aimed to identify opportunities and barriers to the smart transformation of the steel industry in Industry 4.0 conditions. This segment was established by two questions: Q7: What factors are opportunities of industrial transformation of steel enterprises to

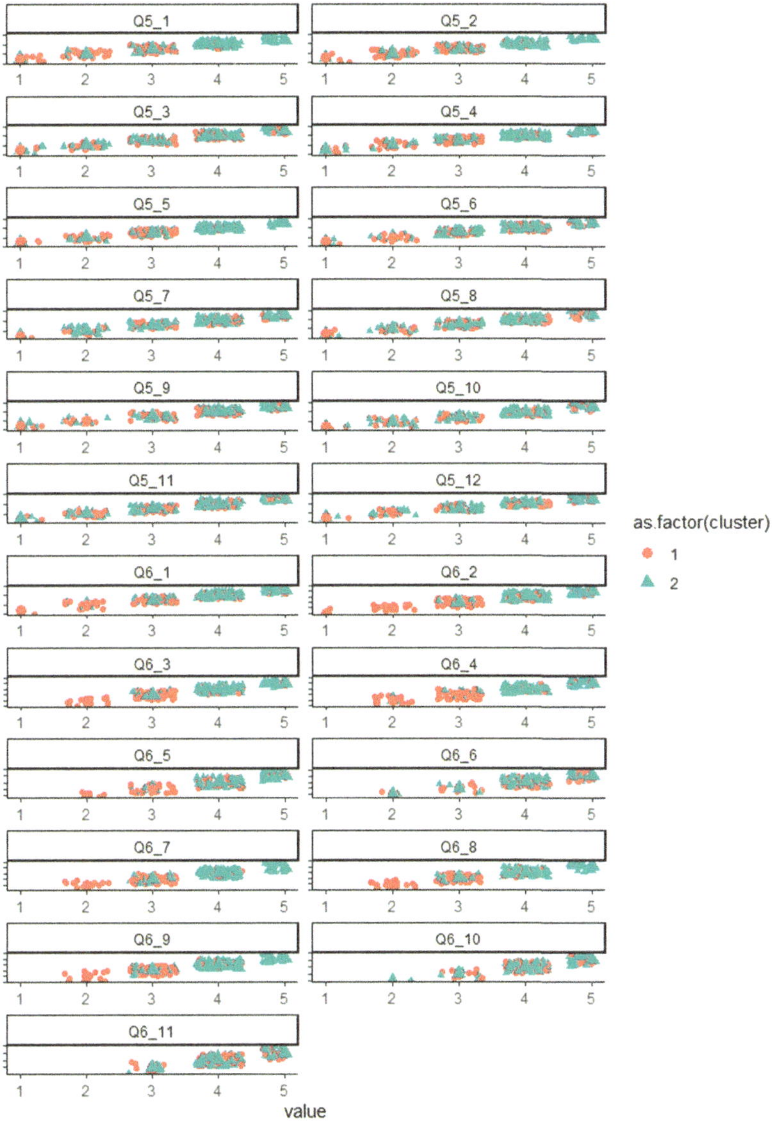

Fig. 4.15 Discrepancy analysis in the research area: RA_3. Source: own research

the Industry 4.0? Q8: What factors are barriers to industrial transformation of steel enterprises to Industry 4.0? Table 4.11 presents the structure of responses according to the Likert scale.

Based on the structure of the share of responses (Table 4.11) for ranks 4 and 5, the following opportunities were identified: (Q7_6) development of customer services (steel products are accompanied by

Table 4.11 Structure of response shares in the RA_4 based on the questions: Q7 and Q8

Opportunities and barriers of industrial transformation				
Question	*2*	*3*	*4*	*5*
Q7_1. Research and development/product design/ service diversification	2.4%	21.7%	44.3%	31.6%
Q7_2. Procurement/delivery/supply chain cooperation	1.2%	15.0%	45.5%	38.3%
Q7_3. Automation of manufacturing processes, optimization of processes	2.4%	20.6%	46.6%	30.4%
Q7_4. Warehousing /distribution/flexibility/ speed/ coperation/resilience	1.2%	16.2%	42.3%	40.3%
Q7_5. Sales and marketing/consumer communication/ consumer cooperation	1.6%	12.3%	49.0%	37.2%
Q7_6. Consumer services /pre- and post-sales/additive manufacturing/product processing	0.8%	9.5%	39.1%	50.6%
Q7_7. Business inteligence and networks	0.4%	17.0%	45.5%	37.2%
Q7_8. Maintenance/servicing/repair/TPM	2.4%	22.1%	45.5%	30.0%
Q7_9. Information management and communication (ICTs/Internet/IIOT)	0.4%	10.7%	39.9%	49.0%
Q8_1. Lack of staff with digital skills	2.4%	21.7%	45.1%	30.8%
Q8_2. High cost of purchasing new equipment/ technology/computer systems	1.6%	25.3%	41.1%	32.0%
Q8_3. Economic calculation/ uncertain return on investment (ROI)	1.2%	13.4%	48.6%	36.8%
Q8_4. Sectoral government policy/environmental fees, environmental investments/energy crisis	2.0%	16.2%	39.5%	42.3%
Q8_5. Competition in the steel market, exports of products from third countries (outside the EU)	0.8%	15.8%	46.6%	36.8%
Q8_6. High costs of reorganization and replacement of steel production technology (industrial zero-emission transformation)	1.2%	17.8%	40.3%	40.7%

Source: own research

No one selected answer 1, which means that the factors were well matched to the characteristics of the I4.0 concept

services, e.g. cutting of products, painting, corrosion protection, extrusion, etc.) Development of the incremental manufacturing market is a strong opportunity for industrial transformation of the steel industry); (Q7_5) marketing and development of sales forms, (Q7_9), development of ICTs and Internet. Among barriers are: economic calculation of technological changes (Q8_3), industrial policy gaps and environmental costs (Q8_4), increasing competition in steel markets from third countries (outside the EU) (Q8_5). Figure 4.16 presents the opportunities and threats of smart transformation based on changes in steel industry in Poland. Based on a total number of responses for ranks 4 and 5 (Fig. 4.16), the analysis revealed that opportunities outweigh the threats for smart transformation. The ranking of barriers is primarily driven by market and cost-related factors (Q8_3)—the steel market is currently in recession due to the impact of the COVID-19 pandemic and the energy crisis, compounded by the high costs of investment in advanced technologies. Additional cost-related (Q8_6, Q8_2) and market-related (Q8_5) factors also contribute to these barriers. The remaining two barriers are related to reorganisation efforts, supported by a strong demand for employees with digital skills (Q8_6, Q8_1).

Steel companies in Poland see opportunities in further technological investments, which should be closely integrated with organisational

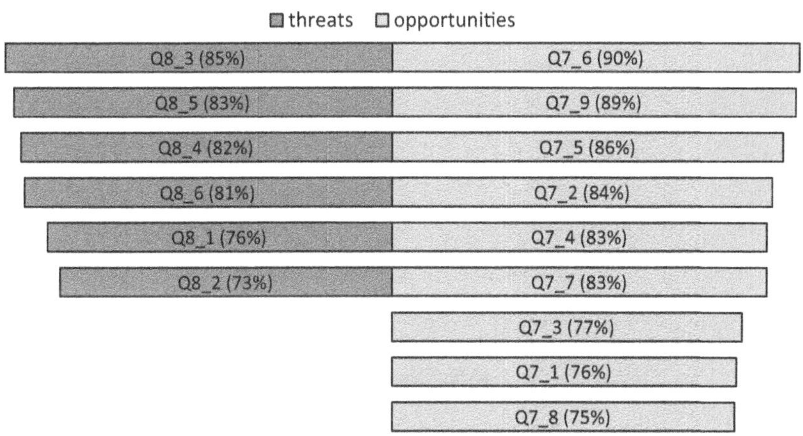

Fig. 4.16 Opportunities and threats of smart industrial transformation in the Polish steel industry. Source: own research

management levels (Q7_9, Q7_7), in collaboration with customer needs research and customer-focused solutions (Q7_6, Q7_5), as well as the supply chain, with particular emphasis on automating the storage and distribution processes within steel operations (Q7_2 and Q7_4).

The last two opportunities are the improvement of maintenance processes (Q7_8) and the enhancement of R&D activities within enterprises (Q7_1). The opportunity to increase R&D activity is primarily targeted at large steel enterprises.

Figure 4.17 presents a vertical analysis of response shares in the RA_ 4/Q7 and Q8. The group "a" presents the following opportunities: customer services market, development of ICTs and the Internet. These segments received the highest ranks on the Likert scale, and the Q7_6 (development of customer market and services) answer had a median = 5.

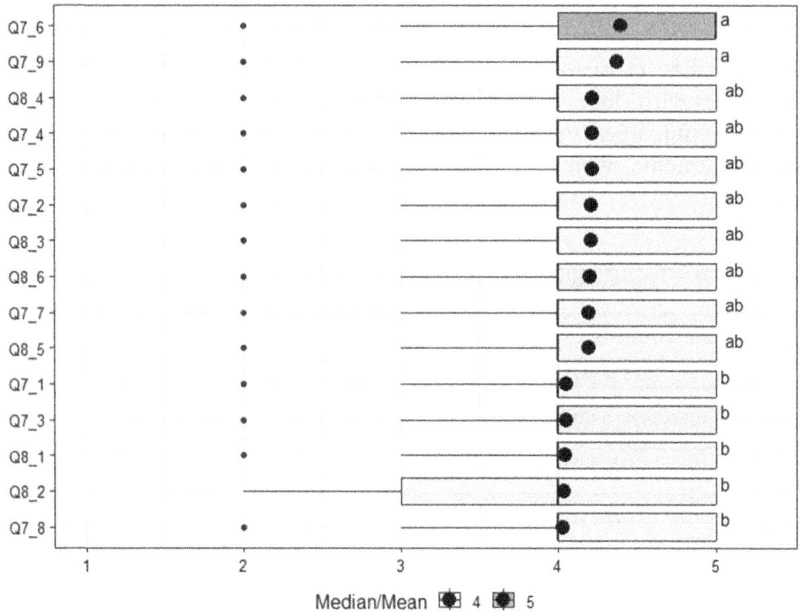

Fig. 4.17 Boxplot of Response Distribution and Median/Mean Comparison Across Questions with Tukey's HSD Test Results in the research area:RA_4. Source: own research

Research Area: RA_5

The last research area (RA_5) was about investment plans and expected progress in industrial transformation of steel enterprises. The segment of research was formed from two questions: Q9. What are investment plans of your enterprise in industrial transformation towards Industry 4.0 ? (responders tick 3 elements/items); Q10: When will there be real benefits from implementing Industry 4.0 technology in your enterprise? (answers to the question: We see benefits now = 5 points (Q10_1), benefits will be in 5 years = 4 points (Q10_2), benefits will be in 10 years = 3 points (Q10_3), benefits will be more than a decade later = 2 points (Q10_4), benefits will be there after more than a decade and maybe even later—1 point (Q10_5), responders tick one answer). The results of the analysis are presented in Table 4.12.

For the Q9 area, the percentages do not sum to 100% due to the use of multiple-choice questions in the planned investment section. The percentages reflect the number of respondents who provided answers, not the total number of responses. The order (list and ranking) of planned investments within the smart transformation framework in the Polish steel industry is presented in Fig. 4.18.

The two main directions of planned actions in the smart transformation of the steel industry in Poland, focus on continuing the implementation of computerised process support systems (Q9_5) and leveraging IT industry services to enhance business digitalisation and employing the external IT specialists (Q9_2).

The next area of investment includes investments in new equipment (Q9_6) and activities aimed at enhancing digital knowledge management (Q9_3). The fourth area of investment is based on improving existing machinery (Q9_7) and developing customer services (Q9_10). Responds suggest that all changes are realised by supporting of strategic level in their enterprises (Q9_9).

Based on the structure of responses for question Q9 (Table 4.12), the following three key groups of investments were identified: development of computer systems in particular levels of process management (Q9_5), development of digital skills (digital skills are needed on the particular levels of organisation) and IT services a (Q9-2, Q9_3, Q9_4), automation of the manufacturing and logistics (intelligent sensors, machines with better productivity parameters, additive manufacturing, autonomous vehicles, maintenance, smart warehouses (Q9_6, Q9_7). According to the respondents, about 2030 the main benefits of the investments already

Table 4.12 Structure of response shares in the RA_5 based on the questions: Q9 and Q10

Investment plans and expected progress in industrial transformation	
Question	*Percentage of respondents*
Q9_1. No activities are carried out or planned in this area	0.0%
Q9_2. Investing in external IT services and support for enhancing business digitalization/ external specialits are recruited	50.2%
Q9_3. Management training/development of digital skills/reorganization	37.9%
Q9_4 Training of employees of selected operational departments/ development of digital skills	21.7%
Q9_5. IT investment/computer systems/integration of IT systems/ horizontal and vertical integration	51.8%
Q9_6. Investing in additive manufacturing, high technology, autonomous vehicles, code readers, smart sensors/clouds (emerging technologies)	36.4%
Q9_7. Investing in machines/machine maintenance/automation of operations/ production lines (existing technologies)	27.7%
Q9_8. Visualization of processes/computerized process control systems/ remote control/autonomous decision-making systems	16.6%
Q9_9. Developing strategic-level initiatives within the organization towards smart transformation	30.0%
Q9_10. Customer services, personalisation, castomisation, interoperable technologies	27.7%
Q10_1. We see benefits now (5)	21.3%
Q10_2. Benefits in 5 years (4)	23.3%
Q10_3. Benefits in 10 years (3)	19.4%
Q10_4. Benefits in 10 + years (2)	18.2%
Q10_5. Benefits step by step, continuous process (1)	17.8%

Source: own research

being made (ROI) should be visible. Figure 4.19 presents the changes on the timeline (progress of changes).

Description of the figure: time from 2020 because the study was started in 2020.

4.3 Research Conclusions

With regard to the adopted study areas (RAs), the following was established:

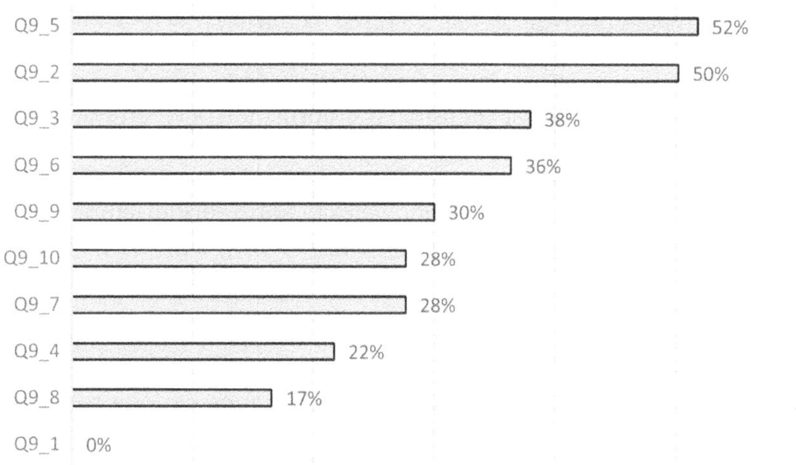

Fig. 4.18 Planned directions of smart transformation in steel industry. Source: own research

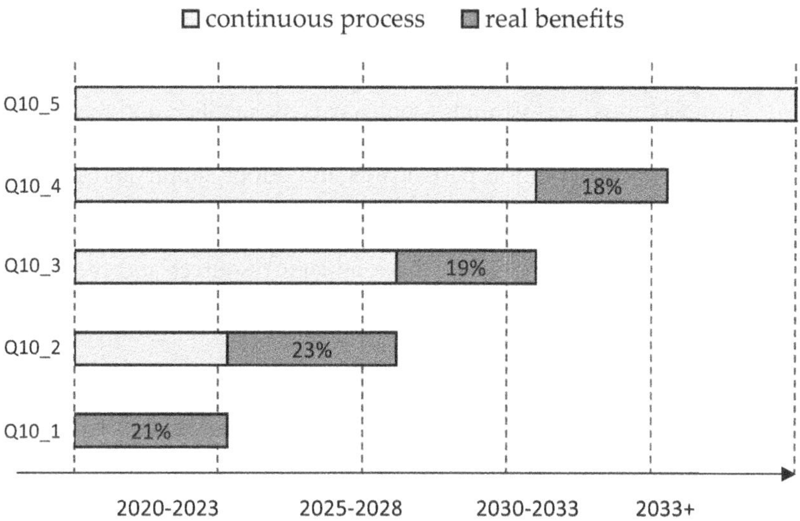

Fig. 4.19 Changes on the timeline: progress of changes in the smart transformation of the Polish steel industry. Source: own research

- The representatives of the surveyed companies were knowledgeable about the concept of Industry 4.0 and its technological pillars. Their understanding of the idea of Industry 4.0 proved sufficient to conduct studies on the industrial transformation of enterprises in the era of Industry 4.0.
- The surveyed companies are investing in high productivity and precision manufacturing technologies (additive manufacturing) and computer-supported process systems (ERP, MES, WMS, CRM, CAD, among others), with the availability of cloud services and support from the IT sector.
- Information technology and the Internet support the collection and processing of data needed to optimise manufacturing and logistics processes. Furthermore, information and communication technologies (ICT), such as EDI systems, e-invoicing and product ordering through company websites, have facilitated customer service. In focusing on customer orientation, the surveyed companies have placed an emphasis on continuous improvement of customer service processes, using mobile technology and the Internet to strengthen customer relationships and personalise products.
- Industry 4.0 technologies, according to the respondents, can significantly improve operations within companies and supply chains. Steel manufacturers are strongly connected to supplier and consumer markets, thus they strive to integrate the information systems of individual collaboration participants and adopt supply and product tracking technologies (blockchain technology). Blockchain can be used to track the history of raw materials from extraction to production. In the steel industry, where mineral resources and metals are crucial, blockchain enables the creation of an immutable and transparent history of a given piece of metal. This assures buyers of its origin, whether it comes from legal sources, whether forced labour was involved and its properties. Blockchain enables better supply chain management in the steel industry. By recording transactions and the movement of goods on blockchain technology, companies can track every step, from extraction through processing to the delivery of the final product. This increases transparency in collaboration, reduces the risk of errors and streamlines logistics.
- In the coming years, planned investments will continue to focus on data processing technologies and improving communication, as well

as enhancing processes within companies, primarily through horizontal and vertical integration. New technologies (AI) require IoT, meaning a network of interconnected devices (M2M).
- Companies pointed out that technological transformation must be carried out alongside reorganisation. The steel industry increasingly needs employees with digital skills and more IT services. The surveyed companies employed external staff to carry out individual investment projects, particularly during the design phase and the synchronisation of IT systems, sensorisation of devices, and the creation of technological platforms with access to cloud solutions and the Industrial Internet of Things (IIoT).
- On their path to Industry 4.0, respondents identified barriers belonging to economic factors (ROI, investment costs) and market-political factors (market competition, environmental costs and others).

4.4 Limitation of the Research and Future Study

The study carried out had some limitations. Firstly, the study was exploratory and its aim was to obtain general knowledge of the transformation of steel enterprises in the Polish steel market, but without details about technological solutions applied in steel mills and enterprises cooperating with them. Realised questionnaire was designed to address the problem of the transformation of steelworks to Industry 4.0 under conditions of overriding environmental policies, and the questions were constructed on the basis of accepted research assumptions according to the nine pillars of Industry 4.0, which are the general requirements of this concept. The survey lacked detailed information on the technologies implemented, such as the types of technologies. In future research, the author will aim to establish details about the smart steel transformation.

Secondly, the research was realised and information was collected in the period from 2020 to 2023. The time horizon was long, and during that time there were market fluctuations in the steel market, including the negative effects of the COVID-19 pandemic, the war in Ukraine and the energy crisis. These distortions may have affected the value of the study and investment process in steel mills (Gajdzik et al., 2024).

The selection of enterprises in the survey was deliberate, and the respondents were people who were able to answer questions about the ongoing transformation and at the same time worked in independent positions in the smelters as well as in companies with associated smelters. The target selection did not include enterprises that are just at the beginning of their journey in the transformation underway, as well as enterprises that are just planning their first investments. Therefore, the target selection of enterprises for research could be a limitation for other enterprises, that is a third limitation.

For the purposes of this book, research results are grouped into clusters of change, while the range of statistical correlations between the scope of technological changes and types of enterprises is much wider. Many statistical correlations have been analysed, but are not presented in this book. The author intends to prepare scientific publications based on them. Moreover, the level of maturity of digital transformation in the Polish steel industry was separately analysed and presented in the other book (see: Gajdzik et al. book, 2025a). The leader of changes in the Polish market is ArcelorMittal Poland: As one of the largest steel producers in Poland, ArcelorMittal has implemented a range of smart manufacturing technologies, including IoT sensors and AI-driven analytics, to optimise production and reduce emissions. The case study was presented in the book: Gajdzik et al., 2025b).

The author is also aware of the scope of research focused on one topic. Changes in steel companies are affected by a number of factors that were not established during the investigation. The author's future research will focus on the relationship between "smart manufacturing" and "green transformation." The strategic direction of the transformation of steel mills to operate in Industry 4.0 is part of a broader strategy of change, which is sustainable development. Therefore, the ongoing industrial transformation, with strong support from digital and smart technologies that are transforming steel manufacturing into smart steel manufacturing, will be increasingly focused on "green and smart steel manufacturing." The key future research: "green" changes is steel transformation towards smart steel manufacturing.

References

Büchi, G., Cugno, M., & Castagnoli, R. (2020). Smart factory performance and Industry 4.0. *Technological Forecasting & Social Change, 150*, 11979. https://doi.org/10.1016/j.techfore.2019.119790

Castelo-Branco, I., Cruz-Jesus, F., & Oliveira, T. (2019). Assessing Industry 4.0 readiness in manufacturing: Evidence for the European Union. *Computer in Industry, 107*, 22–32.

Dubey, G., Gupta, R. K., Kumar, S., & Kumar, M. (2022). Study of industry 4.0 pillars and their uses in increasing productivity and reducing logistics defects. *Materials Today Proceedings, 63*, 85–91. https://doi.org/10.1016/j.matpr.2022.02.335

Erboz, G. (2017). How to define Industry 4.0: Main pillars of Industry 4.0. In *7th international conference on management (ICoM 2017)*, At Nitra, Slovakia, November 2017. Retrieved June 9, 2019, from https://www.researchgate.net/publication/326557388_How_To_Define_Industry_40_Main_Pillars_Of_Industry_40

Gajdzik B. (2022) *Diagnoza kierunków transformacji przemysłu stalowego w Przemyśle 4.0* (*Diagnosis of the steel Industry transformation direction in the Industry 4.0*). Silesian University of Technology.

Gajdzik, B., Jaciow, M., Wolniak, R., & Wolny, R. (2025a). *Maturity business models for manufacturing in the digital age*. CRC Press, Boca Ration, Taylor & Francis Group, London - New York (A Science Publisher Book). ISBN: 978-1-032-72970-1 (hbk), ISBN: 978-1-032-74168-0 (pbk), ISBN: 978-1-003-46795-3 (ebk). https://doi.org/10.1201/9781003467953

Gajdzik, B., & Wolniak, R. (2021). Digitalisation and innovation in the steel industry in Poland—Selected tools of ICT in an analysis of statistical data and a case study. *Energies, 14*, 3034. https://doi.org/10.3390/en1411303410.3390/en14113034

Gajdzik, B., Wolniak, R., Grebski, W., Szymszal, J., & Grebski, E. M. (2025b). *Sustainability and smart manufacturing. The transformation of the steelwork industry*. Routledge.

Gajdzik, B., Wolniak, R., Nagaj, R., Žuromskaitė-Nagaj, B., & Grebski, W. W. (2024). The influence of the global energy crisis on energy efficiency: A comprehensive analysis. *Energies, 17*, 947. https://doi.org/10.3390/en17040947

Lee, E. A. (2008). Cyber physical systems: Design challenges. In: *11th IEEE symposium on object oriented real-time distributed computing (ISORC)* (pp. 363–369).

Machin, D., Campbell, M. J., Tan, S. B., & Tan, S. H. (2009). *Sample size tables for clinical studies* (3rd ed.). Wiley-Blackwell.

Peters H. (2017). How could Industry 4.0 transform the Steel Industry? Presentation at Future Steel Forum, Warsaw, 14–15.6.2017. Retrieved from https://futuresteelforum.com/content-images/speakers/Prof.-Dr-Harald-Peters-Industry-4.0-transform-the-steel-industry.pdf

Rojko, A. (2017). Industry 4.0 concept: Background and overview. *International Journal of Interactive Mobile Technologies, 11*(5), 7. https://doi.org/10.3991/ijim.v11i5.7072

Rüßmann, M., Lorenz, M., Gerbert, P., Waldner, M., Justus, J., Engel, P., & Harnisch, M. (2015). *Industry 4.0*. The Boston Consulting Group.

Senn, C. (2019). The nine pillars of Industry 4.0. Retrieved from https://www.idashboards.com/blog/2019/07/31/the-pillars-of-industry-4-0/

CHAPTER 5

A Model Framework for Smart Steel Transformation

Abstract This chapter, which summarises both the literature and empirical parts of the analysis, presents the constructs of smart steel manufacturing model, which is positioned in a sectoral environment that will change under the influence of the decarbonisation of industries, including the steel industry and technological trends of Industry 4.0. The concept of Industry 4.0 is the current vision of industrial development, shaping the future of many industries, including metallurgy. Technological pillars of this concept create new business models based on solutions of smart manufacturing. Therefore, it is crucial to investigate its dimensions in order to realise efficiencies in the business environment. In this book, the industrial trends are shortly described as well as how they would help companies achieve their objectives concerning Industry 4.0 in sustainable development. Key industry trends frame changes in business models. Many modern trends are strongly oriented towards Industry 4.0, but the technological changes companies are undertaking must contribute to achieving sustainable development.

Keywords Business model · Sectoral surrounding · Smart steel transformation · Industry 4.0

B. Gajdzik, *Digital Transformation towards Smart Steel Manufacturing*, https://doi.org/10.1007/978-3-031-77756-1_5

5.1 Model Framework of Smart Steel Manufacturing

Recent industrial technological changes are centred around the concept (idea) of Industry 4.0. This idea is a combination of technologies, devices, and processes integrated into the production procedures in the entire supply chain and value chains, promoting self-sufficient production processes and decentralised decisions requiring minimum human intervention. This compilation is based on a strong integration (symbiosis) between digitalisation and executing technologies. According to Hofmann and Rüsch (2017), this integration is possible due to consistent and flexible connectivity and computerisation, which permits a higher level of automation and self-adapting production systems and thus enables autonomous and decentralised decision-making (Kagermann et al., 2011). Innovative solutions are based on the Internet, IIoT, blockchain, Big Data, cloud computing, robotics, human–machine interaction, artificial intelligence, open source software and others. All these solutions are based on digitisation, which is supported by new capabilities of process and communication people and machine and machine and machine (Davies, 2015). According to the assumptions of the Industry 4.0 concept, companies strive to compile digital and operating technologies into cyber-physical production systems (Lee, 2015; Lee et al., 2015a, 2015b; Liu et al., 2017).

Industrial changes based on the operating technologies create solutions for smart manufacturing. According to Kan et al. (2016), smart manufacturing is the collection of cutting-edge technologies that support effective and accurate engineering decision-making in real time through the introduction of various ICTs and the convergence with the existing manufacturing technologies. Changes are focused on manufacturing from the smart factory perspective (Osterrieder et al., 2020). Future manufacturing is becoming "smart"—capable of agilely adapting to a wide variety of changing conditions in Industry 4.0. Future smart factories will be at the centre of every business network (Lu & Ju, 2017).

The smart factory is based on an architecture consisting of four layers (Osterrieder et al., 2020 The first, the physical layer, collects data through sensors and interconnected physical machines. The second and third are responsible for sending data from these machines to the cloud, where the data is stored, processed and analysed. In the latter, the control layer, a program that can be monitored by staff, controls the entire smart factory.

The four layers follow the four design principles described by Hermann et al. (2016) interconnectivity, information transparency, decentralised decisions and technical support. Digitalisation of the production process and artificial intelligence are determining the directions of industrial developments and are placing it in the architecture of a new business model (Zhong, et al., 2017).

Investments in industries targeting highly advanced technologies that have a strong impact on competitive conditions in the global marketplace and directions of governmental policies. In the happening revolution, the availability of new technologies is a strong incentive to develop and improve production systems, in which many advantages are seen in the scale of economies. Industry 4.0 is a set of disruptive technologies changing how enterprises in all industries plan, perform and evaluate their business strategies (Culot et al., 2020).

Changes in business are strongly linked in economic, social and ecological trends. According to Muslemani et al. (2021), decarbonisation in line with the Paris Agreement's target of reducing global net emissions to zero by 2050 (UNFCCC, 2015) is a major challenge for industries. This trend is embedded in the balanced development of current and future economies. Decarbonisation technologies can permanently change industries. Therefore, it is necessary to include deep decarbonisation trend into the smart manufacturing model.

The new business model should include both solutions that promote smart manufacturing and green manufacturing (Fig. 5.1). The final results of industrial transformation will be smart and green products (in the steel sector, the names "green steel" and "smart steel") (Vogl et al., 2020). A number of existing initiatives address issues relevant to the two directions of development, for a green steel market such as standardisation of product environmental footprint calculations and responsible steel (ResponsibleSteel, 2019), regulatory framework for CO_2-lean (Eurofer, 2019), energy efficiency (Gajdzik et al., 2021a, 2021b; Wolniak et al., 2020), low-carbon manufacturing (2023b; Gajdzik et al., 2023a), business model development for cleaner technologies in steelmaking (Muslemani et al., 2020) and for a smart steel market such as computer systems for process optimisation and quality control (Miśkieiwcz & Wolniak, 2020), predictive maintenance (Lee et al., 2015a, 2015b), fully automatic with machine learning, autonomous robots and other intelligent applications (Rudy et al., 2018; Zhou et al., 2022), system for autonomous decision-making (Rodriguez et al., 2021), blockchain technology (Jenabi,

2023) and others. Consistent climate and environmental protection and a successful smart manufacturing are not mutually exclusive, in fact, they need each other.

Industrial transformation cannot be achieved by any single steel mill. Advancing industrial transformation requires interaction between industry partners and various institutions and organisations. Innovation ecosystem partners include steel mills, material and energy suppliers, distributors (steel product warehouses), steel product consumers, government agencies, business associations, universities and industrial research and development centers. The innovation ecosystem is a construct of many actors and many activities, in global, national, regional and local scopes (Matt, et al., 2021; Oh, et al., 2016).

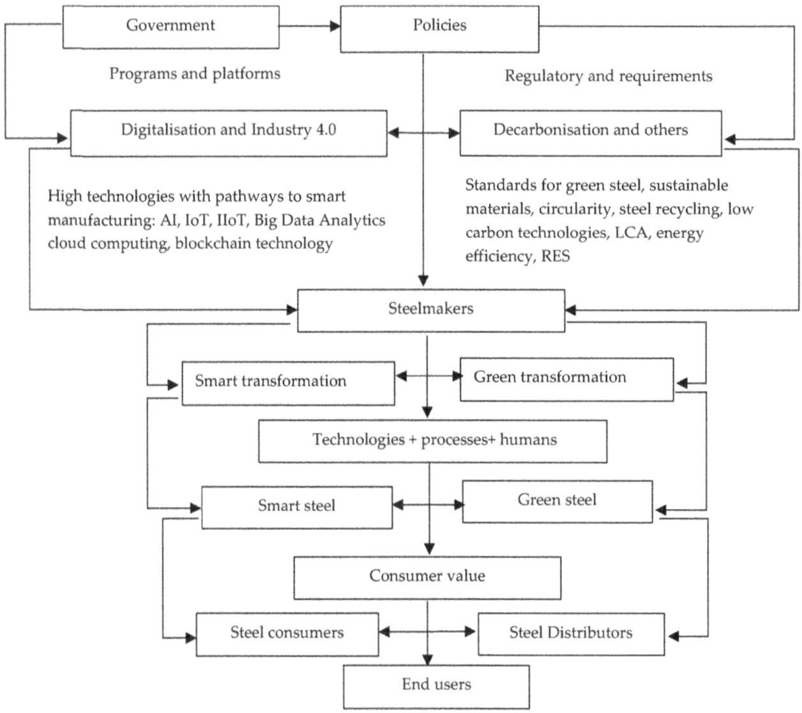

Fig. 5.1 Model framework of steel industry transformation towards "smart" and "green' steel. Source: own elaboration

Transforming steel manufacturers to Industry 4.0 needs flexible and resilient supply chains. Steel producers are strongly linked to suppliers of materials, energy and services and production volumes depend on demand in the markets of steel consumers. Cooperation between companies in the steel industry is based on a number of dependent and independent relationships. In this marker, there is a strong dependence of steelmakers on steel customers. It can be assumed that the strategy of smart manufacturing in steel mills is a form of response to customer needs. The transformation in steel industry needs concurrent manufacturing technologies and real-time sharing of process data. Digitisation of supply chains supported by high technologies is a determinant of modern steel value chains in the global steel market. Strategical directions of changes are determined by the technologies of Industry 4.0, the forms of cooperation and connections in the supply chain, as well as the environment with social, economic trends and governmental policies (Fig. 5.2).

In Industry 4.0, innovative technologies offer a unique set of innovations that can revolutionise the steel sector. The steel industry can harness vast amounts of data to optimise operations. Predictive maintenance using data analytics reduces downtime and increases machinery reliability. Real-time data analysis improves product quality by monitoring and controlling variables, while demand forecasting ensures a seamless production process aligned with market trends. Robots and AI systems are becoming integral to steel manufacturing. Autonomous robots handle material handling and logistics, reducing labour costs and enhancing efficiency. AI-driven predictive maintenance optimises machinery performance and safety, with robotic systems often deployed for hazardous tasks (Dai & Zuo, 2020; Gajdzik, 2014).

The steel industry is continuously evolving to embrace the transformative potential of Industry 4.0. Companies in this sector are implementing new technologies to create and operate in a smart environment. Changes are implemented gradually, from simple station projects to large technological investments (Gajdzik & Wolniak, 2021b; Gajdzik et al., 2021a, 2021b). In the steel industry transformation, the distinctions between digitisation and smartness are blurring. Changes in both areas are interpenetrating and creating solutions that in the future can be considered smart steel manufacturing. Innovations are implemented in the surveyed companies on a continuous, albeit segmented (piecemeal) basis on selected processes or technology islands (Gajdzik, 2021, 2022).

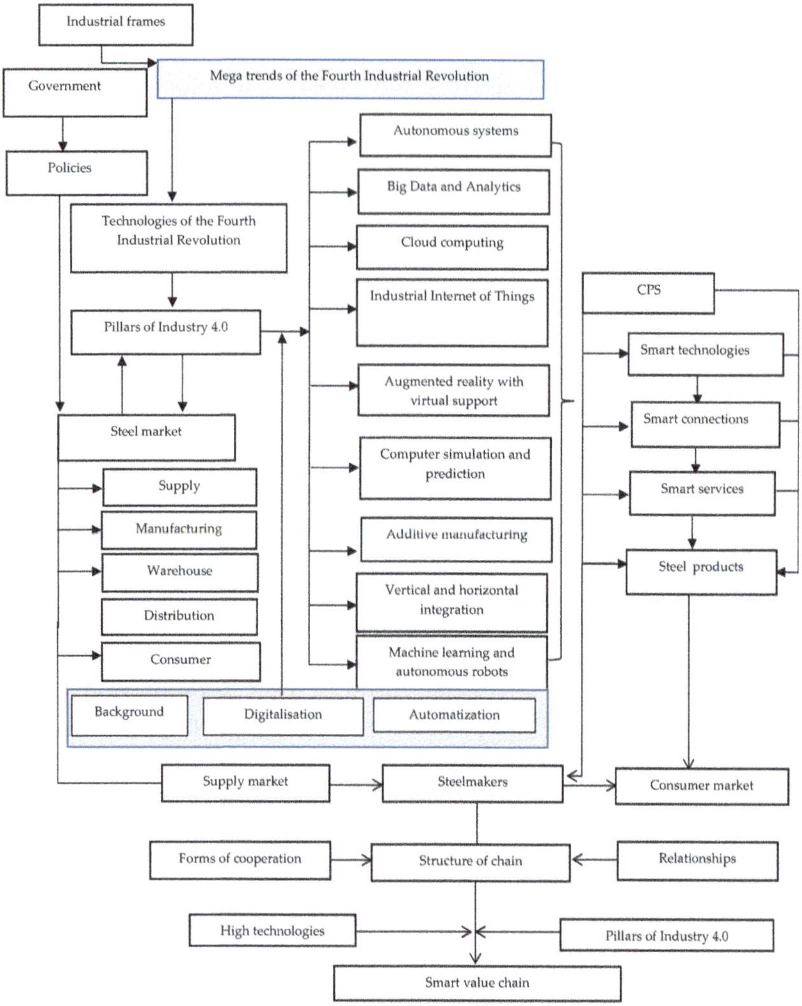

Fig. 5.2 Model framework of steel transformation in the era of Industry 4.0. Source: own elaboration

Digitisation has entered the metallurgy gradually, covering subsequent processes. Digital technologies are used, among others, to monitor the operation of devices in selected sections of production, technological maintenance services (MS), control of parameters of the production environment, control of equipment operation and tracking of processes. IT process support systems are considered the beginning (base) of the transformation. Steel companies, for years, have been betting on the synchronisation of IT process support systems (ERP, MES, WMS), and in recent years there have been increased investment in intelligent sensors that inform about the operations and processes in real time. The automation of manufacturing operations is still developing (Danel & Gajdzik, 2024; Danel et al, 2023; Herzog, 2017).

Integrated Management Information Systems (IMISs) are modularly organised IT systems that support all areas of enterprise activity, from marketing, planning and supply, through technical preparation of production, management of the manufacturing process, distribution, sales, repair management, to financial accounting work and human resources management (Danel & Gajdzik, 2024). Integration within steel mills includes three levels:

- strategic (management) control level,
- tactical (dispatching) control level,
- operational control level.

The strategic (management) control level is usually allocated to the company management level and is connected with long-term planning, balance reporting and production/sales analysis. The centre of tactical control of the preparation plant is the control room, dispatching, loading control or shift foremen's worksite. Users have at their disposal information about the progress of the controlled process. The operational control level is related directly to a technological process. Human participation is limited to supervision and setting of control parameters. This is primarily the direct control of technological nodes in distributed control units and automated data collection. Process control takes place in real time. With the development of ICTs, control systems have access to Big Data and cloud computing. Automatic identification of physical objects on the network is done using RFID technology. With the help of sensors, it is possible to automatically detect the status of a device, read out process parameters in real time, transmit control signals, exchange messages and cooperate with other devices connected via wireless networks (Danel & Gajdzik, 2024). Without IoT, the devices deployed in steel mills would not be able to create smart solutions, and the collected data would not be stored on internal drives, so cloud services are needed. For many steel mills, migration to the cloud is directly related to data modernisation and maintenance. Technologies woven into IIoT and M2M communication with increasingly strong autonomous decision-making processes are moving towards smart steel manufacturing (Danel et al, 2023; Gajdzik, 2014; Gajdzik & Wolniak, 2021a; Cheng et al., 2020).

The expansion of the information system for continuous production (steel mills) should include the following modules and extensions (Danel & Gajdzik, 2024):

- Possibility of connecting sensors and actuators (controllers) via OPC (Open Process Control) industrial buses;
- Support for data collection using IIoT (Industrial Internet of Things) modules, including the possibility of wireless communication. IoT enables the connection of distributed systems and topologically separated workstations;
- Support for material transport monitoring using RFID technology mainly concerns the identification of wagons or materials on conveyor belts and containers;
- Requirement of availability of production balances and overview of the current state of production processes online using web applications;
- Analysis of the amount of data leads to the conclusion that there is no need to implement big data technology;
- Deploy cloud services to simplify system maintenance;
- Supplementing the system with predictive functions using neural networks or artificial intelligence.

Process control of steel manufacturing is transforming from static to dynamic and anticipatory control. Process Control (PC) is a process-oriented method and has the character of active control, as its results are used to identify process disturbances that need to be eliminated. Process control is carried out in real time and its objective is to continuously improve the course of processes and achieve the best results. Modern technologies—process control systems—are highly autonomous, i.e. they are able to control the process without human intervention and make decisions for them (Ahuett-Garza & Kurfess, 2018; Benotsmane et al., 2018).

In the transformation model of the steel industry, the technologies of the pillars of Industry 4.0 intersect and form the basis for cyber-physical systems with strong collaboration in the supply chain. The constructs of cyber-physical systems are smart technologies based on the Industry 4.0 pillars, smart connections with the Internet and IoT and smart services. In the transformation of the steel industry to Industry 4.0, the boundary between products and services is blurring, which is being realized at a higher level of interaction with steel consumers. Steel products are seen as platforms for the provision of whole-system services at the stage of steel product design, processing and coating of products, production of structural components and other manufacturing processes, as well as throughout the logistics processes.

5.2 Key Directions of Smart Steel Transformation

The determination of the directions of changes in the steel industry was completed with a comparison in the context of the development based on technologies of Industry 4. 0 with a strong influence of government policies towards the sustainability of steel production and "green steel". In accordance with the idea of Industry 4. 0 will continue to be implemented:

- machine learning to simplify steelmaking processes and streamline steelmaking operations (computerised process optimisation),
- virtual reality (VR) in business and supply chains backed by blockchain technology for materials and product verification and full product life cycle (LCA) tracking;
- linking smart technologies, cooperating companies, into cyber-physical systems, from raw material extraction to steel production to distribution and sale of finished products,

- automation of operations in processes and improvement of steelmaking technologies towards sustainability—environmental efficiency of steelmakers together with efficiency of steel products throughout their life cycle,
- improving steel products in an effort to produce "smart" products, which are able to provide expected value to customers in dependent markets (construction, mechanical equipment, automotive industry, metal products, other transportation, electric equipment, domestic appliances)
- improving decision-making systems, coping with the challenges of big data and the need to make quick decisions using technologies that create opportunities to build flexibility into steelmaking processes,
- investing in technologies replaces the work of people in hard work conditions, in digital systems, workers are operators of innovative technologies,
- improving smart technologies with machine-to-machine (M2M) communication in steel supply chains to strengthen business network structures while increasing their flexibility (fast response to market needs, shorter delivery times, customisation of steel (metal) products, etc.),
- building the foundation of smart steel mill architecture and end-to-end chain engineering in supply chain structures using advanced and robotic steelmaking technologies and connecting all logistics processes.

The listed directions of technological innovation that make up the smart steel manufacturing model are subject to the regulations of overarching policies, including:

- deep decarbonisation with zero-carbon steelmaking technologies and carbon dioxide utilisation systems (CCUS),
- efficient energy management and diversification of energy sources with a significant share of green energy and green hydrogen,
- direct iron reduction—DRI technology in steel manufacturing,
- recycling of steel scrap and its use in steel production (EAF technology),
- carbon footprint reporting (CBAM) by steel product importers,

- circular economy and LCA supported by blockchain technology.

The listed directions are selected from among many because they are now considered key, but the issues of the relationship of each enterprise, with the environment, are multifaceted, it is impossible to discuss all the scopes of influence, especially in terms of the development of Industry 4.0, which is a technologically and organisationally complex concept.

The overall conclusion: in the era of Industry 4.0 and Industry 5.0, the transformation processes of the steel industry in the global market must be strongly linked to "green" changes in accordance with the requirements of global climate policy. The basis of the ongoing industrial transformation is the relationship: "smart" and "green".

References

Ahuett-Garza, H., & Kurfess, T. (2018). A brief discussion on the trends of habilitating technologies for Industry 4.0 and smart manufacturing. *Manufacturing Letters, 15*, 60–63. https://doi.org/10.1016/j.mfglet.2018.02.011

Benotsmane, R., Dudás, L., & Kovács, G. (2018). The concept of autonomous systems in Industry 4.0. *Advanced Logistic Systems, 12*(1), 77–87. https://doi.org/10.32971/als.2019.006

Cheng, J., & Westman, J. (2020). Effects of digitalization in steel industry economic impacts & investment model. Master of Science Thesis TRITA-ITM-EX 2020:280 KTH Industrial Engineering and Management Industrial Management, Stockholm (pp. 35–39).

Culot, G., Orzes, G., Sartor, M., & Nassimbeni, G. (2020). The future of manufacturing: A Delphi-based scenario analysis on Industry 4.0. *Technological Forecasting and Social Change, 157*, 120092. https://doi.org/10.1016/j.techfore.2020.120092

Dai, H., & Zuo, X. (2020). Industry 4.0 boosts the steel industry into a new era of digitalization and smart technology. *Aistech - Iron and Steel Technology Conference Proceedings, 3*, 1829–1841.

Davies, R. (2015). *Industry 4.0 digitalisation for productivity and growth.* European Parliament PE 568.337, Eur. Parliamentary Res. Service, vol. 1. https://www.europarl.europa.eu/RegData/etudes/BRIE/2015/568337/EPRS_BRI(2015)568337_EN.pdf

Danel, R., & Gajdzik, B. (2024). Integrated subsystems of materials and information flow for continuous manufacturing of coal and steel. *Management Systems in Production Engineering, 32*(1), 174–184.

Danel, R., Gajdzik, B., & Ropyak, L. (2023). Trends in data processing control in continuous production systems. *Organization and Management Series, 181*, 93–105. https://doi.org/10.29119/1641

Eurofer. (2019). A regulatory framework for CO2-lean steel produced in Europe. https://www.eurofer.eu/publications/position-papers/a-regulatory-framework-for-co2-lean-steel-produced-in-europe/

Gajdzik, B. (2014). Autonomous and professional maintenance in metallurgical enterprise as activities within total productive maintenance. *Metalurgija, 53*(2), 269–272.

Gajdzik, B. (2021). Transformation from Steelworks 3.0 to Steelworks 4.0: Key technologies of Industry 4.0 and their usefulness for polish steelworks in direct research. *European Research Studies Journal, 24*(3), 61–71.

Gajdzik, B. (2022). Key directions in changes from steelworks 3.0 to steelworks 4.0 with analysis of selected technologies of digitalizing the steel industry in Poland. *Management Systems in Production Engineering, 30*(1), 46–53. https://doi.org/10.2478/mspe-2022-0006

Gajdzik, B., Grabowska, S., & Saniuk, S. (2021a). A theoretical framework for Industry 4.0 and its implementation with selected practical schedules. *Energies, MDPI, 14*(4), 1–24. https://doi.org/10.3390/en14040940

Gajdzik, B., Sroka, W., & Vveinhardt, J. (2021b). Energy intensity of steel manufactured utilising EAF technology as a function of investments made: The case of the steel industry in Poland. *Energies, 14*, 5152. https://doi.org/10.3390/en14165152

Gajdzik, B., & Wolniak, R. (2021a). Digitalisation and innovation in the steel industry in Poland—Selected tools of ICT in an analysis of statistical data and a case study. *Energies, 14*(334), 2021. https://doi.org/10.3390/en1411 3034

Gajdzik, B., & Wolniak, R. (2021b). Transitioning of steel producers to the steelworks 4.0—Literature review with case studies. *Energies, 14*, 4109. https://doi.org/10.3390/en14144109

Gajdzik, B., Sujová, E., & Biały, W. (2023a). Decarbonisation of the steel industry: Theoretical and practical approaches with analysis of the situation in the steel sector in Poland. *Acta Montanistica Slovaca, 28*(3), 621–636.

Gajdzik, B., Wolniak, R., & Grebski, W. (2023b). Process of transformation to net zero steelmaking: Decarbonisation scenarios based on the analysis of the Polish steel industry. *Energies, 16*, 3384. https://doi.org/10.3390/en1608 3384

Hermann, M., Pentek, T., & Otto, B. (2016). Design principles for Industrie 4.0 scenarios. In *Proceedings of the annual Hawaii international conference on system sciences* (pp. 3928–3937). https://doi.org/10.1109/HICSS.2016.488

Herzog, K. W. (2017). The digitalization of steel production. *BHM Berg- und Hüttenmännische Monatshefte, 162*(11), 504–513.

Hofmann, E., & Rüsch, M. (2017). Industry 4.0 and the current status as well as future prospects on logistics. *Computers in Industry, 89*, 23–34. https://doi.org/10.1016/j.compind.2017.04.002

Jenabi, H. (2023). Revolutionizing the iron and steel industry with blockchain technology. Retrieved from https://www.linkedin.com/pulse/revolutionizing-iron-steel-industry-blockchain-hamed-jenabi

Kang, H. S., Lee, J. Y., Choi, S. S., Kim, H., Park, J. H., Ji, Y. S., Kim, B. H., & Noh, S. D. (2016). Smart manufacturing: Past research, present findings, and future directions. *International Journal of Precision Engineering and Manufacturing-Green Technology, 3*(1), 111–128. https://doi.org/10.1007/s40684-016-0015-5

Kagermann, H., Wahlster, W., Helbig, J. (Eds.). (2011). Recommendations for implementing the strategic initiative Industrie 4.0: Final report of the Industrie 4.0. Working Group: Industrie 4.0: Mit dem Internet der Dinge auf dem Weg zur 4. industriellen Revolution, VDI-Nachrichten, April 2011; Acatech-National Academy of Science and Engineering: München, Germany. http://forschungsunion.de/pdf/industrie_4_0_final_report.pdf

Lee, J. (2015). Smart factory systems. *Informatik Spektrum, 38*, 230–235.

Lee, J., Ardakani, H. D., Yang, S., & Bagheri, B. (2015a). Industrial big data analytics and cyber-physical systems for future maintenance & service innovation. *Proceedings of CIRP, 38*, 3.

Lee, J., Bagheri, B., & Kao, H. (2015b). Research letters: A cyber-physical systems architecture for Industry 4.0-based manufacturing systems. *Manufacturing Letters, 3*, 18–23.

Liu, Y., Peng, Y., Wang, B., Yao, S., & Liu, Z. (2017). Review on cyber-physical systems. *IEEE/CAA Journal of Automatica Sinica, 4*, 27–40. https://doi.org/10.1109/jas.2017.7510349

Lu, Y., & Ju, F. (2017). Smart manufacturing systems based on cyber-physical manufacturing services (CPMS). *IFAC International Federation of Automatic Control, 50–1*, 15883–15889.

Matt, D., Bolzano, I., Molinaro, M., Orzes, G., & Pedrini, G. (2021). The role of innovation ecosystems in Industry 4.0 adoption. *Journal of Manufacturing Technology Management, 32*(9), 369–395.

Miśkiewicz, R., & Wolniak, R. (2020). Practical application of the Industry 4.0 concept in a steel company. *Sustainability, 12*(14), 5776.

Muslemani, H., Liang, X., Kaesehage, K., Ascui, F., & Wilson, J. (2021). Opportunities and challenges for decarbonizing steel production by creating markets for 'green steel' products. *Journal of Cleaner Production, 315*, 128127. https://doi.org/10.1016/j.jclepro.2021.128127

Oh, D. S., Phillips, F., Park, S., & Lee, E. (2016). Innovation ecosystems: A critical examination. *Technovation, 54*, 1–6.

Osterrieder, P., Budde, L., & Friedli, T. (2020). The smart factory as a key construct of industry 4.0: A systematic literature review. *International Journal of Production Economics, 221*, 107476. https://doi.org/10.1016/j.ijpe.2019.08.01

ResponsibleSteel Standard v1.0. (2019). ResponsibleSteel website at www.responsiblesteel.org/standard-development/

Rodriguez, J., Kothiyal, M., Kalvenes, J., Wolfgang, M., Lukic, V., & Nath, G. (2021). Strengthening the steel industry with AI. BCG publication. Retrieved from https://www.bcg.com/publications/2021/value-of-ai-in-steel-industry

Rudy, V., Rusnák, J., & Mares, A. (2018). Application of Industry 4.0 tools in metallurgy processes. *Transfer Innovation. Kosice TU, 38*, 57–59.

UNFCCC. (2015). Adoption of the Paris Agreement FCCC/CP/2015/10/Add.

Wolniak, R., Saniuk, S., Grabowska, S., & Gajdzik, B. (2020). Identification of energy efficiency trends in the context of the development of Industry 4.0 using the Polish steel sector as an example. *Energies, 13*, 2867. https://doi.org/10.3390/en13112867

Vogl, V., Åhman, M., & Nilsson, L. J. (2020). The making of green steel in the EU: a policy evaluation for the early commercialization phase. *Climate Policy, 21*, 78–92.

Zhong, R. Y., et al. (2017). Intelligent manufacturing in the context of Industry 4.0. *Engineering, 3*(5), 613–630.

Zhou, D., Xu, K., Lv, Z., Yang, J., Li, M., He, F., & Xu, G. (2022). Intelligent manufacturing technology in the steel industry of China: A review. *Sensors, 22*(21), 8194. https://doi.org/10.3390/s22218194

CHAPTER 6

Summary

Keywords Changes · Digital transformation · Smart steel manufacturing

Digital transformation (DT) is reshaping industries worldwide, including the steel manufacturing sector. For Polish steel enterprises, the journey towards "smart manufacturing" involves adopting advanced technologies that improve efficiency, sustainability and competitiveness. This framework outlines key elements to guide steel companies in Poland through the digital transformation process, creating a blueprint for a smart steel manufacturing future.

6.1 Key Drivers for Digital Transformation in Steel Manufacturing

Global Competition: As steel production costs rise, the Polish steelmakers must improve technological processes to stay competitive.

Sustainability and Regulations: Growing environmental regulations require the Polish steel plants to reduce carbon emissions and black energy consumption, prompting the need for green steel and decarbonisation of steelmaking processes.

B. Gajdzik, *Digital Transformation towards Smart Steel Manufacturing*, https://doi.org/10.1007/978-3-031-77756-1_6

Efficiency and Productivity: Automation and data-driven processes can significantly enhance production, reduce downtime and improve overall operational efficiency. Industry 4.0 technologies are needed in improving steelmaking processes in companies and their chains.

6.2 Technological Pillars of Smart Steel Manufacturing

Internet of Things (IoT): Connected devices and sensors can collect real-time data throughout the steel production process, from raw material handling to final product delivery. This enhances decision-making and predictive maintenance.

Artificial Intelligence (AI) and Machine Learning (ML): AI can optimize production processes by analysing massive datasets and predicting machine failure, while ML improves product quality and reduces waste as well as energy consumption.

Big Data Analytics: By analysing production data, steel enterprises can identify patterns, optimise processes and make informed business decisions. Access to the data improves computer simulations of process flows and the implementation of operational changes. Computer systems: Computer systems to support steel production processes are integrated and autonomous.

Robotics and Automation: Automated systems and autonomous technologies (machines) can handle repetitive tasks in steel manufacturing, improving precision and reducing human error. Additive Manufacturing: new solutions (technologies, such as 3D printers) produce new products. Steel mills are also using 3D printers in internal maintenance. Augmented Reality: physical objects and virtual models create solutions that streamline many functions and operations. Augmented reality technology is useful in the design and assembly of components, and can also promote workplace safety.

Cloud Computing: The cloud enables seamless data storage, sharing and processing, allowing steel companies to operate more flexibly and scale operations as needed.

Cybersecurity: With increased digitisation comes a higher risk of cyber threats. Polish steel manufacturers must prioritise robust cybersecurity measures to protect sensitive data and systems.

6.3 Challenges to Digital Transformation in the Polish Steel Enterprises

Computer Systems: Steel plants still improve systems that may easily integrate with modern technologies.Digitization of processes, automation of operations and other technologies embedded in the pillars of Industry 4.0 have entered steel mills.

Skills Gap: A shortage of workers skilled in digital technologies can slow the adoption of smart manufacturing solutions.

Digital Pathways: Steel companies need IT industry support at the stage of synchronizing systems and technologies.The transformation of steel mills to Industry 4.0 is staged, and its scope varies, at the level of the individual pillars of Industry 4.0, as well as the expected results.

High Initial Investment: Although digital transformation promises long-term cost savings, the upfront investment in technology and infrastructure can be a barrier for many enterprises.

Continuous Improvement: Digital transformation is not a one-time process but an ongoing journey. Regularly assess progress and make adjustments to improve efficiency and competitiveness.

6.4 Conclusion

For the Polish steel enterprises, digital transformation (DT) is not just an option but a necessity in today's competitive and environmentally conscious market. By adopting smart manufacturing technologies, companies can improve operational efficiency, reduce environmental impact and maintain their competitiveness on a global scale. A well-structured framework with clear goals, technology adoption, workforce development and strategic partnerships will be critical in driving this transformation forward.

This framework provides a roadmap for steel enterprises in Poland, emphasising that the path to smart steel manufacturing is not just about technology but also about a shift in mindset, skills, and collaboration.

Index

A

additive manufacturing (AM), 3, 5, 12, 15, 30, 38, 39, 69, 84, 86, 88, 99, 117, 120

artificial intelligence (AI), v, 3–5, 11, 13–16, 28, 30, 33, 34, 38, 95, 126, 127, 132, 140

augmented and virtual reality, 28

automation, v, vi, 2, 4, 5, 7, 9–11, 30, 31, 37–39, 57, 67, 96, 101, 108, 112, 117, 126, 131, 134, 140

B

Basic Oxygen Furnace (BOF), 28, 36, 57

Big Data, 3, 15, 16

Big Data Analytics (BDA), 3, 5, 8, 10, 12, 31, 90, 140

Blast Furnace - Basic Oxygen Furnace (BF-BOF), 42, 58, 64, 74

Blast Furnace—Basic Oxygen Furnace (BF-BOF), 42, 44

blockchain, 5, 15, 16, 31, 39, 40, 96, 100, 105, 120, 126, 127, 133, 135

C

carbon, 3, 16, 43, 45, 74, 127, 139

carbon dioxide (CO_2), vi, 16, 36, 41, 43, 44, 46, 74

challenges, 15, 16, 37, 40, 127, 134, 141

changes, v, vi, 2–5, 16, 32, 35–37, 41, 42, 44, 55, 57, 72, 74, 81, 82, 95, 102, 105, 108, 112, 115, 117–119, 122, 126, 127, 129, 133, 135

cloud computing (CC), 3, 4, 9, 12, 31, 33, 86, 89, 90, 100, 126, 131, 140

cluster analysis, 82, 90, 94, 99, 102

collaboration, 2, 11, 116, 120, 133, 141

computer, 5, 8–10, 14, 17, 28, 29, 31, 32, 35, 57, 67, 82, 87–90,

B. Gajdzik, *Digital Transformation towards Smart Steel Manufacturing*, https://doi.org/10.1007/978-3-031-77756-1

92, 96, 99, 101, 102, 105, 109, 117, 127, 141
consumer, 8, 9, 13, 15, 28, 30, 35, 40, 44, 48, 58, 62, 99, 100, 128, 129, 133
consumption, 3, 4, 16, 28, 58, 61, 62, 64, 66, 69, 74, 139
control, 3, 13, 28, 29, 31–33, 35–37, 56, 67, 96, 109, 126, 131–133
cyber-physical systems (CPS), v, 3, 4, 7, 11, 34, 75, 88, 91
cybersecurity, 6, 13, 88, 99, 104, 140

D
decarbonisation, vi, 36, 41, 43, 45, 46, 74, 75, 127
decarbonising, 46, 47
deep decarbonisation, vi, 3, 5, 41–43, 46, 127
digital technologies, v, 2, 6, 8, 28, 31, 35, 37, 75, 131, 141
digital transformation (DT), v, vi, 4, 6, 11, 13, 16, 122, 139, 141
digitisation, v, vi, 2, 4–7, 10, 31, 32, 126, 129
direct reduction of iron (DRI), 42, 47, 74, 134
direct reduction of iron-electric arc furnace (DRI-EAF), 42, 47
direct research, vi, 80, 84
distribution, 5–7, 38, 40, 72, 86, 89, 90, 100, 108, 112, 116, 131, 133

E
economy, v, 2, 4, 5, 13, 15, 16, 46, 48, 55, 58, 64, 69, 72, 127, 135
efficiency, v, 4, 7, 11, 16, 18, 28, 36, 37, 42, 64, 67, 127, 134, 140, 141
electric arc furnace (EAF), 28, 42, 46, 58, 70, 74, 134
employees, 9, 17, 57, 64, 115, 121
energy, v, 3, 5, 7, 16, 17, 28, 35, 41–44, 46, 64, 66, 67, 70, 74, 115, 121, 127–129, 134, 139
energy intensity, 16, 30, 32, 64, 66, 74
Enterprise Resource Planning (ERP), 7, 8, 14, 31, 32, 37, 40, 105, 120, 131
European steel industry, 33, 41, 72

F
flexibility, v, 33, 34, 38, 40, 81, 109, 112, 134

G
green steel, 41, 43–46, 48, 74, 127, 133

H
hydrogen, 41, 42, 44–47, 66, 74, 134

I
Industrial Internet of Things (IIoT), 3, 5, 10, 11, 31, 89, 90, 99, 100, 104, 121, 126, 132
industrial transformation, 10, 18, 28, 75, 80, 82, 95, 99, 105, 112, 115, 117, 122, 127, 128, 135
Industry 4.0, v, vi, 3–5, 10–12, 14, 17, 18, 57, 69, 75, 80, 83, 84, 89–91, 93, 95, 102, 104, 108, 112, 114, 117, 120–122, 126, 127, 129, 130, 135
Industry 5.0, vi, 4, 13, 15–18, 37, 135
information and communication technologies (ICT), v, 3–5, 31,

37, 67, 105, 115, 116, 120, 126, 131
integration, 4, 13, 28, 33, 34, 36, 37, 84, 91, 96, 102, 105, 126, 131
Internet of Everything (IoE), 2, 5, 8, 12
Internet of Things (IoT), v, 2, 3, 5, 12, 14–16, 33, 38, 86, 140
investment, 34, 41, 44, 57, 66, 67, 70, 75, 112, 115, 117, 120, 121, 127, 129, 141

L
learning machine (LM), 30
life cycle, 4, 16, 37, 40, 134

M
machine, 7, 11–13, 29, 126, 140
maintenance, 7, 15, 17, 28–30, 36, 39, 69, 117, 132, 140
management, 6–8, 16–18, 28–30, 32, 35–40, 69, 95, 108, 112, 116, 117, 120, 131, 134
model framework, vi, 128, 130

N
network, 7, 9, 12–14, 16, 17, 30, 31, 34, 35, 38–40, 72, 100, 121, 131, 132, 134

O
operator, 11, 36, 57, 134
optimisation, 28, 29, 90, 109

P
pillars, 11–13, 17, 69, 83, 84, 89, 93, 95, 104, 133
pillars of Industry 4.0, vi, 3, 12, 80, 84, 85, 95, 105, 121, 133
Poland, vi, 55–64, 66–75, 80–82, 102, 112, 115, 117, 122, 139, 141
policy, v, 2, 5, 36, 43, 44, 57, 75, 115, 121, 127, 129, 133–135
Polish steel industry, 55–58, 64, 66, 67, 69–72, 75, 80, 95, 115, 117, 119, 122
Polish steel market, vi, 56, 58, 65, 66, 71, 80, 121
presonalisation, 109
process, v, vi, 2, 4–13, 17, 18, 28–32, 34–39, 41–43, 47, 57, 58, 60, 64, 66, 67, 69, 71, 75, 83, 86, 88–90, 92, 96, 99, 102, 105, 108, 109, 112, 116, 117, 121, 126, 127, 129, 131–133, 135, 140, 141
processes integration, 108, 112
producers, 40, 43, 44, 56, 58, 71, 74, 75, 122, 129
product, v, 2, 6, 12, 13, 15, 16, 29–32, 34–36, 38–40, 43, 45, 46, 62, 64, 69, 70, 73, 108, 109, 114, 120, 127, 128, 133, 134, 140
productivity, 5, 11, 16, 29, 30, 33, 34, 57, 64, 109, 112, 117, 140

Q
quality, v, 2, 11, 13, 29, 30, 35, 36, 104, 108, 109, 112, 127, 129, 140

R
resilience, 4, 13, 17, 37, 112
respondents, vi, 82–86, 89–91, 93, 95, 99, 102, 112, 117, 121
restructuring, 34, 56–58, 64, 71, 74
robotization, v, vi, 2, 4, 10, 30, 96, 108

robots, 8, 10, 11, 16, 17, 39, 89, 93, 96, 99, 101, 102, 127, 129

S

sensors, 4, 7, 9, 12, 16, 29, 30, 32, 33, 37, 38, 67, 96, 105, 117, 126, 131, 132, 140
smart manufacturing (SM), v, 5, 12, 13, 30, 33–35, 80, 99, 101, 102, 105, 122, 126–129, 139, 141
smart steel manufacturing, vi, 28, 33–35, 80–82, 99, 122, 126, 129, 132, 134, 139–141
smart steel transformation, 121, 133
steel, v, 2, 28, 31–33, 36, 40, 42, 44, 47, 55–58, 62–64, 67, 69–72, 74, 75, 81, 84, 99, 112, 116, 120, 122, 127, 129, 134, 140
steel companies, vi, 56, 72, 96, 102, 105, 115, 122, 131, 139–141
steel enterprises, vi, 80, 82, 84, 95, 99, 102, 112, 114, 116, 117, 139–141
steel industry, vi, 28, 33–35, 38, 42–44, 46, 47, 57, 63, 64, 69, 71, 82, 112, 115, 121, 129, 133, 135
steelmaking, v, 28, 32, 38, 41, 43, 44, 47, 66, 74, 127, 133, 134
steel manufacturing, 30, 32, 35, 44, 122, 132, 139
steel market, vi, 38, 74, 80, 115, 127
steel mills, vi, 28–30, 32, 34, 40, 42–44, 56, 57, 64, 66, 67, 69, 71, 72, 74, 81, 121, 122, 128, 129, 131, 132, 134
steel process, 30, 34, 35
steel production, 30–34, 41, 46, 56–59, 64, 67, 69, 72, 74, 133, 134, 140
steel products, 30–32, 35, 40, 44, 56, 58, 62, 63, 69, 72, 73, 108, 128, 133, 134
steelworks, 28, 31, 56, 58, 70, 74, 121
supply chain (SC), v, vi, 4–6, 12, 16, 28, 30, 31, 35, 37–40, 46, 69, 75, 81, 105, 108, 109, 112, 126, 129, 133, 134
sustainable development, 122
systems, v, 7–9, 12, 13, 16, 28–32, 34, 37, 39, 40, 57, 67, 86, 89, 92, 96, 99, 100, 102, 105, 109, 117, 127, 129, 131–133, 140, 141

T

technological pillars, 11, 80, 95, 120, 140
3D printing, 3, 4, 12, 15, 30, 32, 69, 84, 86, 88, 99, 101, 104
trade, 30, 62, 63, 69
trends, v, 2, 11, 38, 62–64, 75, 112, 127, 129

V

vertical and horizontal analysis, 95
vertical and horizontal integration, 2, 7, 12, 28
visualisation, 36, 38

The manufacturer's authorised representative in the EU is Springer Nature Customer Service Centre GmbH, Europaplatz 3, 69115 Heidelberg, Germany. If you have any concerns regarding our products, please contact ProductSafety@springernature.com

Printed and bound by CPI Group (UK) Ltd, Croydon, CR0 4YY
30/09/2026
02225472-0008